RESISTING EVICTION

RESISTING EVICTION

Domicide and the Financialization of Rental Housing

Andrew Crosby

Fernwood Publishing
Halifax & Winnipeg

Copyright © 2023 Andrew Crosby

All rights reserved. No part of this book may be reproduced or transmitted in any form by any means without permission in writing from the publisher, except by a reviewer, who may quote brief passages in a review.

Copyediting: Erin Seatter
Cover design: Evan Marnoch
Printed and bound in Canada

Published by Fernwood Publishing
2970 Oxford Street, Halifax, Nova Scotia, B3L 2W4
and 748 Broadway Avenue, Winnipeg, Manitoba, R3G 0X3
www.fernwoodpublishing.ca

Fernwood Publishing Company Limited gratefully acknowledges the financial support of the Government of Canada through the Canada Book Fund and the Canada Council for the Arts. We acknowledge the Province of Manitoba for support through the Manitoba Publishers Marketing Assistance Program and the Book Publishing Tax Credit. We acknowledge the Nova Scotia Department of Communities, Culture and Heritage for support through the Publishers Assistance Fund.

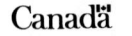

Library and Archives Canada Cataloguing in Publication
Title: Resisting eviction : domicide and the financialization of rental housing / Andrew Crosby.
Names: Crosby, Andrew A., 1979- author.
Description: Includes bibliographical references and index.
Identifiers: Canadiana (print) 2023046114X | Canadiana (ebook) 20230463061 | ISBN 9781773636375 (softcover) | ISBN 9781773636511 (EPUB) | ISBN 9781773636528 (PDF)
Subjects: LCSH: Rental housing—Ontario—Ottawa—Case studies. | LCSH: Eviction—Ontario—Ottawa—Case studies. | LCSH: Landlord and tenant—Ontario—Ottawa—Case studies. | LCSH: Low-income tenants—Ontario—Ottawa—Case studies. | LCGFT: Case studies.
Classification: LCC HD7288.85.C22 O83 2023 | DDC 363.509713/84—dc23

Contents

1 | Revitalization and Settler Colonial "Improvement" 1
    Manufactured Housing Crises 4
    Demovicting and Defending Heron Gate 6
    Ottawa: The "Liveable" City 9
    Book Structure 10

2 | The Racial Logics of Property 14
    Settling and Unsettling Ottawa 15
    Tools of Dispossession 16
    The Reproduction of Settler Colonial Urbanism and Racial Capitalism 18
    Reverberations of Empire: The Coloniality of Migration and Settler States 22

3 | Domicide in the Liveable City 24
    Urban Liveability 24
    Domicide and the Unmaking of Home 27
    The Reproductive Side of Domicide 29
    Gentrification and Settler Vitality 31

4 | Research Methods and Design 35
    Institutional Ethnography 36
    Political Activist Ethnography 37
    Movement-Relevant Theory and Activist Scholarship 38
    Data Collection and Analysis 40

5 | Heron Gate, Racial Stigma, and Strategic Neglect 48
    A Liveable Community? 48
    Racial Stigma 56
    Strategic Neglect 59

6 | Heron Gate and the Financialization of Rental Housing 65
    The Ownership Trajectory of Heron Gate 65
    Housing Policy, Corporate Capture, and Financialized Gentrification 67
    "The Apartment as Saviour": Strategies of Real Estate Investment Firms 70
    Timbercreek: "Actively Creating Value" 72

7 | Demoviction 2016: Domicide and Redevelopment in Heron Gate 80

Revitalization and Relocation 80
Intensification and Demoviction 82
The "Bombing of Heron Gate" 84
Vista Local: Revitalization and Resort-Style Living 85

8 | Demoviction 2018: Tenant Resistance to Domicide 96

Unwilling Subjects of Financialized Gentrification 97
Public Shaming, Social Media, and Legal Repression 104
Flipping Defamation: Suing the Landlord 111
Landlords: "How to Handle a Crisis" 113

9 | Community Wellbeing: A Social Framework for Domicide 115

Manufacturing Consent for the Master Plan 116
A Social Framework for Heron Gate 119
"A Termination Plan": The Heron Gate Official Plan Amendment 125

10 | Human Rights and Racial Discrimination in Housing 142

A Precedent for Housing Rights 142
The Strategic Value of Legal Engagement 146
Housing Futures: Resisting Domicide in the Liveable City 147

Acknowledgements 149

References 151

Index 169

This book is dedicated to the wonderful people that have been involved with the Herongate Tenant Coalition. They welcomed me into their struggle with open arms and I have learned so much from them all. I hope the words in the following pages do justice to their work and can be useful in the ongoing struggle as well as future tenant movements. The Herongate Tenant Coalition's uncompromising determination, resolve, and commitment to tenant and housing justice is a source of infinite inspiration and hope.

CHAPTER 1

Revitalization and Settler Colonial "Improvement"

I currently reside as an uninvited guest on unceded Algonquin Anishinaabe land, in what settler society refers to as the Ottawa-Gatineau or national capital region of Canada, but I was born and raised on an island in the heart of the Mi'kmaw homeland of Mi'kma'ki. Epekwitk, formerly known as Île-Saint-Jean, is now commonly referred to as Prince Edward Island. When I was a teenager living with my mom and sister, we resided in a section of the north end of Summerside on Epekwitk. The neighbourhood is composed predominantly of rowhouses built in the 1960s at the height of the Cold War to house military personnel that worked at the Canadian Forces Base. When the base closed in 1991 and military families vacated, the rowhouses offered affordable living for residents in the economically depressed town. The homes were large, with two storeys, a basement, and multiple bedrooms; they had plenty of living space. The make-up of the neighbourhood offered a sense of community, with lots of green space and room for children to play and thrive. I never felt a sense of stigma living there, even though in hindsight I see it was considered lower-income housing. When I moved to Ottawa and was introduced to the broader Herongate and Heatherington neighbourhoods, I noticed a remarkable similarity in the look and feel there compared to where I had spent my formative teenage years.

When I first left home, I moved into various apartment rental units in Summerside before eventually leaving the island altogether, renting in cities such as Halifax and Ottawa. My mom also moved around, a lot. She left Epekwitk, or Prince Edward Island, for Nova Scotia, before settling into New Brunswick (all of these so-called Maritime provinces are part of Mi'kma'ki, which stretches beyond imposed provincial and federal borders to encompass parts of Newfoundland, Quebec, and Maine). My mom and stepfather moved into an apartment complex in Dieppe (adjacent to Moncton) in 2007. This became our family's first experience living in a complex owned by a large corporation that would end up selling the property to a financialized landlord. Actually, this apartment complex changed ownership three times in less than two years, and my mom's experience there represents an anecdotal microcosm of what tens of thousands of renters across Canada have likely experienced as their homes are flipped to, and become the property of, real estate investment firms.

My mom's apartment building in Dieppe was first sold around 2012 to a company called MetCap Living, a Toronto-based property management company boasting over $2 billion in assets and owning over twenty thousand rental units in Canada. (MetCap Living gained notoriety among housing and tenant rights activists in Toronto as a wealthy and abusive landlord and was the target of a successful rent strike in the city's Parkdale neighbourhood in 2017.) At the time, the rent in Dieppe was relatively affordable, around $650 for a good-sized two-bedroom apartment, heat and lights included. However, amid declining maintenance levels and increasing rents, MetCap sold the property to TransGlobe in 2012.

This is where my connection to Heron Gate — the Ottawa community at the centre of the story in this book — becomes uncanny. Around that time, TransGlobe gained international notoriety under Daniel Drimmer as one of the leading slumlords in Canada (CBC 2012), embracing an enrichment strategy that at its core embodied purposeful neglect. By 2012 TransGlobe had to strip itself of its image. It morphed into True North Apartment Real Estate Investment Trust and sold its Heron Gate holdings to Timbercreek Asset Management (which would also have to rebrand in the coming years as Hazelview). A version of the company maintains ownership over the Dieppe property to this day.

These connections are both eye-opening and common in the world of real estate investment capital. Yet there is more to this story that links my family's personal experiences with the seemingly inevitable encounter with property ownership. My mom fell into a cycle of renting cheaper units in poorly maintained low-rise apartment buildings. Displeased with the deteriorating living conditions at her apartment in Dieppe, my mom moved to Sackville, New Brunswick, a few dozen kilometres east down the Trans-Canada Highway. Discouraged with the conditions in her first building there, she moved into an old, drafty house that had been converted into apartment units. Sadly, this house burned to the ground a few months later, with my mom and stepfather lucky to make it out with minor injuries, as the unit had no working smoke detectors or fire extinguishers. Overnight, they lost all of their worldly possessions and were rendered homeless. Due to their situation, they were able to secure a tiny, subsidized apartment in a housing complex for seniors that includes small four-unit cottages surrounding a long-term care home facility. At the time I viewed this as a blessing, as the affordable rental units were highly sought after with a years-long waitlist. However, after a few years the low quality of living that the unit offered was evident, with poor air quality and cramped living conditions. My mom wanted out, but the rental market was persistently spiralling out of reach for low-income households.

My mom was desperate to move but could not find even a half-decent place within her budget. This desperation prompted me to begin scouting the local housing market for an opportunity to purchase a house. I was a bit resigned that this seemed to be the only viable option to get her into an adequate living situation. To own property in Canada, to own a piece of Indigenous land, is one of the most sought-out aspects of settler society and considered a pinnacle of achievement. Homeownership is a settler norm that has been driven for decades by federal housing policy. Despite the ethical violations involved, it appears nearly unavoidable. For example, most people I know, including some on the furthest left of the political spectrum, either own property or are seeking to get into the game. And what a game it is, engaging in relationships with bankers and lawyers and a lifetime of debt just to receive a certified piece of paper validating ownership of a very small piece of Indigenous land. To cut to the chase, I found a very nice bungalow that was completely modified for seniors' living and spent my doctoral scholarship savings on a down payment. I later added my mom to the deed and we are now joint tenants of this home, according to the legal terminology. After years of precarious living as a renter, my mom now enjoys decent and stable living conditions, made possible only through homeownership in a settler society that necessitates contributing to the ongoing theft and occupation of Indigenous land.

Manufactured Housing Crises

This experience — apartment transactions within the financialization of rental housing, limited access to affordable and adequate rental housing, and personal engagement in homeownership — transpires amid what is widely considered a housing crisis. Housing in Canada is now defined by multiple forms of manufactured crises in relation to shelter access and affordability, with public discourse predominantly fixating on rising house prices and the increasing inability of settlers and immigrants to realize the Canadian dream of homeownership. Featuring less prominently are the crises of evictions and rental housing insecurity, conditions precipitated by the current housing system, which is animated by rentier capitalism and the predominant emphasis on homeownership. The manufactured crises of the current housing system inevitably lead to domicide, the deliberate destruction of home (Porteous and Smith 2001), through gentrification, urban development, eviction, and displacement. Yet domicide does not go uncontested. Investigating private, high-end housing development and mass demolition-driven eviction (or demoviction) as a form of domicide, this book engages with resistance to domicide at a key site of urban revitalization in the capital city of the Canadian settler colonial state.

The steady rise of eviction rates around the globe has been described as "the silent social tsunami of our times" (Soederberg 2018, 286). Rental housing insecurity affects millions of urban households, and interdisciplinary research has attended to accelerated processes of gentrification-induced displacement and the urgent need to decommodify and defend housing (Desmond 2016; Madden and Marcuse 2016; Soederberg 2021). The lucrative political economy of evictions and rental housing insecurity is driven by housing financialization, privatization, and a broader neoliberal assault on housing policy (Brickell, Arrigoitia, and Vasudevan 2017; Sassen 2014; Watt 2018). At the same time, displacement is driven by general trends of gentrification and removal of affordable housing options, where previously affordable dwellings are either transformed or torn down and lower-income tenants are replaced by tenants that can afford higher rents. These trends disproportionately impact marginalized populations along race, gender, and class lines.

Radical approaches to social, political, economic, and environmental justice provoke deeper reflection on the meaning and root causes of various crises. Despite broad recognition of multiple crises within the housing field, there is a lack of deep understanding of what this actually means and what can be done about it (Madden and Marcuse 2016). This is evident when various levels of government (like the City of Ottawa or federal government) declare housing and homelessness emergencies and largely ignore their root causes. Instead, access to shelter is viewed as a technical problem that can be solved, typically with market-oriented measures such as policy modifications related to zoning and land use, mechanisms to expand homeownership, and various technological innovations. Madden and Marcuse (2016) instead want us to view housing in terms of a political-economic problem where the housing system is shaped by wider social conflict between different groups and classes. The key conflict shaping housing systems today is "between housing as lived, social space and housing as an instrument for profitmaking — a conflict between housing as home and as real estate" (Madden and Marcuse 2016, 4).

Housing justice activists and tenant organizers that are engaged in struggle to defend their homes and communities see through the veneer of crisis popularized in media and government discourse. A more radical approach questions the idea of "crisis," as the term implies a departure from a norm; in the case of housing, a crisis implies that universal access to adequate and affordable housing is normal and that we are experiencing a temporary dislocation from a well-functioning standard (Madden and Marcuse 2016, 9). However, inadequate and unaffordable housing is a long-standing norm for lower-income and marginalized populations.

Over the course of carrying out the research for this book, I participated in numerous events that featured tenant organizers and housing activists from multiple cities that shared their perspectives, approaches, and best practices in the face of gentrification and eviction. In the majority of these cases, organizers and activists were not engaged at the municipal level to try and negotiate policy tweaks that would allow for the creation of more affordable units. Instead, they were focused on building tenant power through nonhierarchical, community-based organizing, forming tenant committees and other groups that engage in a diversity of tactics in support of and in defence of tenants.

Within these movements and localized struggles, tenant organizers and activists generate their own theories and methods of social struggle, with an acute understanding of how the housing system functions and the forces that shape it. "Housing, precarity and homelessness is not a necessary evil of modern society but is in fact a manufactured crisis that is the direct result of settler colonialist dispossession and the drive of capitalism to expropriate and commodify the necessities of life," said one person at a virtual event where activists were strategizing about a financialized real estate firm's acquisition of a community space in Winnipeg (West Broadway Tenants Committee 2021). The attempted destruction of Ottawa's Heron Gate neighbourhood is emblematic of a manufactured housing crisis, where an affordable, liveable rental community is targeted by a financialized real estate firm for gentrification, and demolished parcel by parcel with the support and approval of municipal actors.

Demovicting and Defending Heron Gate

This book documents demolition-driven evictions and the redevelopment of a racialized rental complex in the City of Ottawa based on my engagement with tenants and community activists, the landlord-developer, and municipal actors over a four-year period (2018–22). In particular, I draw on my work and experience organizing with the Herongate Tenant Coalition in Heron Gate Village, part of the broader neighbourhood of Herongate, which includes social and commercial entities such as schools, shops, and community services (Masuda and Bookman 2018). Throughout this book I use "Heron Gate" to refer to the specific rental complex under corporate ownership and "Herongate" to refer to the broader neighbourhood.

Heron Gate Village, one of a handful of large rental complexes in Ottawa, lies south of Ottawa's downtown core. It includes hundreds of low-density townhouses along with low-, mid-, and high-rise apartment buildings. Prior to the onset of demovictions in 2016, it included some 1,750 households, home to some 4,500 people. From an aerial perspective, Heron Gate resembles a 21-hectare slice of pizza pointing east. The green space of Heron-Walkley Park to the west represents the pizza crust, and Heron Road on top and Walkley Road on the bottom form the sides of the slice. The tip of the pizza is formed where the two thoroughfares converge. The landlord-developer owns most of this chunk of land, save for a small piece in the northwest corner, which is home to the Heron Road Community Centre, and a few other buildings. The easternmost tip of the property is commercial space, once home to the Heron Gate Mall.

I became familiar with the Heron Gate neighbourhood in 2007, when my partner rented a room in the Heatherington neighbourhood to the south. We spent considerable time in the Heron Gate Mall, mostly shopping at the discount grocery store. This mall was a major community hub for Heron Gate residents, especially in winter. They established businesses and had access to culturally relevant goods and services there; the mall was a vibrant economic, cultural, and social space for them. One of Timbercreek's first moves after acquiring this portion of the property in 2012 was to tear down the mall. The building was replaced with box stores reminiscent of settler suburbia, erasing East African and Arab connections to the formal commercial space. This act represented for some the first attempt to rupture existing relations within the neighbourhood, signalling the onset of gentrification and displacement (Xia 2020).

Financialized real estate investment firm Timbercreek purchased the Heron Gate property in 2012 and 2013, and is in the process of demolishing sections of the neighbourhood in order to build over five thousand new units. Two mass evictions in 2016 and 2018 displaced over eight hundred people. The latest set of evictions has been the subject of intense struggle involving the Herongate Tenant Coalition and Timbercreek (now Hazelview). In this book, I use "Timbercreek" or "Hazelview" to refer to the landlord-developer depending on the time frame; I refer to Timbercreek when discussing events prior to the firm's rebranding in November 2020, and Hazelview when discussing events from that point onward. At times I use "Timbercreek/Hazelview" where I think appropriate. The struggle over Heron Gate has played out on various terrains such as the community, the media, and the legal arena, including a potentially precedent-setting case at the Ontario Human Rights Tribunal brought by numerous displaced evictees seeking a right to return.

At the crux of the human rights case is the assertion that the mass displacement of a racialized, immigrant community and elimination of an ethnic enclave violates the right to housing in international law (Yussuf et al. v. Timbercreek 2019). Heron Gate Village is a majority-racialized community home to many immigrants and refugees, including significant numbers of Somali, Arab, and Nepali families. These diasporic communities have ready access to cultural networks, social supports, and amenities. People who have written about the Heron Gate neighbourhood have used different terms to describe its demographic composition. For instance, Xia (2020) calls Heron Gate a "diverse immigrant neighbourhood," while Mensah and Tucker-Simmons (2021) use the designation of "ethno-racial enclave." During my interviews with people who live in the community, I have also heard Heron Gate described as a cultural enclave, an ethnic enclave, and an ethnic neighbourhood.

The first phase of redevelopment involved the eviction of an estimated two hundred people and the demolition of eighty townhomes in 2016, to make way for a new apartment complex called Vista Local. Completed in 2020, Vista Local offers "liveable homes," "resort-style living," and "safe and healthy communities" (City of Ottawa 2018; Shaw 2017). This highly contentious development is designed to align Heron Gate with the largely white, affluent demographics of Alta Vista, an adjacent neighbourhood. The second phase of redevelopment, in 2018, involved the eviction of some six hundred people, 93 percent of whom were racialized and half of whom were of Somali descent, according to a survey conducted by the Herongate Tenant Coalition. An additional 150 townhomes were demolished.

In this book I engage the struggle at Heron Gate from a standpoint of political activist ethnography, a methodological approach that aims to produce knowledge from an activist perspective that is useful for social movement struggles. Political activist ethnography is a form of qualitative inquiry that focuses on work with and for social movements. It emphasizes that activists hold particular social and political insights into the institutions and social forces they struggle against, and ultimately seek to change social relations by investigating and disrupting the organizing logics of ruling relations. Ruling relations emanate from institutions and discourses that manage and thus dominate society (Smith 1990). In the case of Heron Gate, ruling relations include the web of actors and processes involved in gentrifying and redeveloping the neighbourhood. Engaging qualitatively with social movement struggles over urban development in Ottawa, I interrogate ruling relations by examining discourses of urban governance (liveability) and practices of domicide enacted by ruling actors and institutions (such as city officials and planners, developers, and financialized landlords).

The large-scale Heron Gate redevelopment carries with it dramatic implications for gentrification, displacement, and affordable housing; it also provides insight on the racial logics of property relations in settler society. Hundreds of lower-income, racialized tenants are being evicted from affordable dwellings, dislodged from an ethnic neighbourhood, and replaced by tenants with greater purchasing power in the attempted remaking of an entire community. The redevelopment is facilitated, in part, through discourses of improvement surrounding revitalization and liveability.

Ottawa: The "Liveable" City

Shortly after construction commenced on the first demovicted parcel of land in Heron Gate, the City of Ottawa (2021a) approved a New Official Plan with the stated goal of transforming Canada's capital into North America's most "liveable" mid-sized city. This strategic document concerns how the city will grow and how policies will support economic and community development (City of Ottawa 2019a). It is one in a series of urban plans dating back to 1903, but officials consider this latest iteration to be a "milestone plan" as it focuses on the sophistication and maturation of the capital as it evolves into a "world city" (City of Ottawa 2021a, 10). The redevelopment of Heron Gate cannot be disassociated from mechanisms of municipal governance that enable and facilitate gentrification and domicide. The City of Ottawa embarked on a process to update its Official Plan in September 2018, and the new plan received formal approval in October 2021, one month after city council accepted a proposal to demolish 559 more homes in Heron Gate and intensify the property with dozens of new apartment towers.

How can a city declare itself liveable while overseeing large-scale domicide in one of its most racialized neighbourhoods? This question drove me to try and understand how ruling relations governed by private property work to produce liveable (largely white) and disposable (largely racialized) subjects in the development and redevelopment of a settler city on stolen Indigenous land. I closely followed the City of Ottawa's New Official Plan process and participated in numerous community engagement–style events, along the way collecting and analyzing dozens of documents associated with each stage of the project. I also interviewed a number of elected officials and City of Ottawa employees in various units.

The path for Ottawa's transformation is directed by what are referred to as the "five big moves" — growth, mobility, urban design, resiliency, and economy — which "collectively represent the guiding vision en route to Ottawa becoming the most liveable mid-sized city in North America" (City of Ottawa 2021a, 34). They serve as "essential touch points for land use decisions and policy directions" over the twenty-five-year life cycle of the New Official Plan (City of Ottawa 2021a, 13). Cross-cutting issues include intensification, economic development, energy and climate change, healthy and inclusive communities, gender equity, and culture (City of Ottawa 2021a). Intensification, in particular, is identified as a method "of renewal and injecting new life into existing areas of the city" (City of Ottawa 2019b, 1). As an urban planning document focused on renewal and revitalization, the New Official Plan sets out to fix areas of the city that are devoid of vitality, or life, "by allowing new generations of residents to inhabit existing neighbourhoods" (City of Ottawa 2019b, 1). Revitalization efforts are aimed at producing "new life" — or reproducing settler vitality — through the gentrification of urban spaces and the dispersal and replacement of people that live there.

A document unveiling the five big moves for Ottawa acknowledges the city as prosperous and diverse with a rich history. It further acknowledges that the city was "first" home to the Anishinaabe Algonquin Nation, yet it erases Algonquin sovereignty and jurisdiction over the land. Instead, according to the document, "Ottawa has been shaped by the history of Canada" (City of Ottawa 2019a, 1) even though the Algonquins have never ceded or surrendered territory in their homeland and lay claim to portions of Ottawa — including areas around the Ottawa River that encompass Parliament Hill. The city, having been shaped by Canadian history and settler colonialism, is now a space driven by the quest for perpetual growth, the ongoing expropriation of Algonquin land for urban development, and the reconfiguration and improvement of property for settler prosperity and enjoyment.

Book Structure

This book investigates domicide at Heron Gate Village, characterized by eviction and expulsion of tenants, and the resultant replacement of a lower-income, racialized community with affluent, predominantly white people. Implicated here are powerful economic and governance actors, including a multi-billion-dollar real estate investment firm and officials of the most important political city in the Canadian settler colony. Demoviction and domicide are driven by complementary logics oriented around improving land, maximizing profit, and managing populations.

Efforts to create profits and improve land are reinforced by efforts to produce particular kinds of life — productive, more liveable life. The binding threads are race and property, as efforts to revitalize Heron Gate and create a liveable city are deeply shaped by white supremacy.

I begin Chapter 2 by examining the origins and historical evolution of property relations in Canada's national capital region, outlining some of the colonial tools deployed to dispossess Indigenous Peoples of their land and to produce urban settler formations. The role of racialized property relations is explored in relation to settler colonial urbanism, as well as how diasporic space is produced in settler societies.

Then, in Chapter 3, I explore urban liveability and domicide, identifying liveability as an ideological discourse of urban improvement that is mobilized alongside gentrification efforts to unmake the homes of already marginalized populations. Discourses of liveability, in this regard, work to produce domicide, which I reconceptualize to account for the reproductive and repressive elements of home unmaking. Domicide is structured in part through discourses of improvement (e.g., liveability and revitalization) that hierarchically and racially order life and the value of life.

In Chapter 4 I provide a methodological blueprint for doing research with social movements. I describe my work with the Herongate Tenant Coalition as political activist ethnography, which is a form of qualitative inquiry conducted with and for social movements. Political activist ethnography begins from an activist standpoint, emphasizing that activists hold knowledge and insight into the social forces and institutions they are organizing and struggling against. Moving beyond documenting social struggle, this approach seeks to invigorate progressive social change by investigating and disrupting the organizing logics of ruling relations.

Chapter 5 offers a profile of Heron Gate Village. There I delve into statistics on community demographics and housing. With its high concentration of immigrants, refugees, and people of colour, Heron Gate experiences housing inequality, strategic neglect, and structural racism. However, it is a liveable community that offers strong social supports and cultural networks for its racialized and migrant residents. This chapter also provides a remarkable example of how Heron Gate's landlord manoeuvred municipal legal mechanisms to renege on maintenance obligations for the purposes of hastening demolition.

Chapter 6 then examines Heron Gate's owner and landlord in more detail. I take a deep dive into the world of financialized real estate investment within a broader context of housing policy and deregulation to understand the gentrification strategies deployed by these newer types of apartment investors. I conceptualize financialized gentrification in conversation with the various investment strategies and tactics that financialized real estate unleashes in the built environment, including a new trend of intensification. Finally, I look at how Timbercreek Asset Management rebranded as Hazelview Investments.

Chapter 7 examines the first phase of eviction and demolition in Heron Gate in 2016. I analyze some of the discourses surrounding demoviction, such as revitalization, relocation, and intensification. This discussion sets the stage for a deeper examination of the Vista Local redevelopment that replaced the demolished townhomes on the first demovicted parcel of land in the neighbourhood. I unpack the redevelopment's racist underpinnings and overarching goals to harmonize Heron Gate with Alta Vista.

Chapter 8 documents the second phase of demoviction in 2018 and the emergence of the Herongate Tenant Coalition to try and stop the evictions. I explore a number of coalition tactics including tenant mobilization and social media engagement. I also record Timbercreek's responses to tenant mobilizations, which include techniques of legal repression such as threats of lawsuits and attempts to silence the coalition on social media. The ensuing battles provide lessons for other movements on how to strategically engage their adversaries and how landlords are responding to tenant organizing.

Chapter 9 continues a chronological trajectory documenting the wider struggle and events unfolding around the redevelopment proposal for the neighbourhood. I provide a window into the consultation efforts and community meetings about the landlord-developer's "master plan" for redevelopment, which is based on the Conference Board of Canada's Community Wellbeing Framework (see DIALOG 2022). I explore the framework's emphasis on getting investment returns and shaping desirable conduct, before moving into a deeper investigation of the redevelopment proposal itself. The City of Ottawa greenlit the proposal and demolition of 559 more homes in 2021. An accompanying memorandum of understanding (MOU) establishes terms for future displacement and affordability, which in practice will eliminate affordable housing in Heron Gate. The Herongate Tenant Coalition describes this as a framework for social destruction. I further examine the municipal-developer nexus, including the role of municipal actors and their ties to the real estate development industry.

This book concludes with a discussion of the human rights case initiated by Heron Gate tenants evicted in 2018 who are seeking a right to return. The potentially precedent-setting case before the Ontario Human Rights Tribunal, which considers whether the displacement of racialized tenants is a form of discrimination, could curb the gentrifying endeavours of predatory real estate investment while advancing housing rights in Ontario and Canada.

CHAPTER 2

The Racial Logics of Property

Cities have become a key focus of multidisciplinary inquiry, as human migration and urbanization intensify on a global scale. The increased reorganization and settlement of humans into urban environments has had the effect of mobilizing sociologists, geographers, economists, ecologists, architects, urban planners, and engineers to approach urban processes, forms, and experiences as objects of analysis. Urban sociologists contend that city-making is a social process (Amin 2007; Macionis and Parrillo 2016; Parker 2004), where various social actors engage in an array of formal and informal interactions, including settlement, policy, law, planning, development, production, consumption, struggle, and resistance. For Tonkiss (2013, 3), "such a range of actors raises questions about differential rights to make decisions about and interventions in urban environments, and variable claims to use, make and inhabit city spaces." City-making is determined by the interplay between various modes of organization and interaction, where urban space is produced and reproduced through numerous social and economic arrangements and divisions. In other words, the making of cities and the reproduction of space is contested.

This approach to city-making still largely ignores how urban space is produced and reproduced in settler colonial societies, and how city-making and urban development in these societies are enacted and contested on stolen Indigenous land. To better understand the contemporary social and spatial formations of settler colonial urbanism, especially as it affects Heron Gate today, I contextualize contemporary struggles over urban space by examining the historical development and urbanization of Ottawa in the heart of the Algonquin Anishinaabe homeland.

Settling and Unsettling Ottawa

The lands comprising the 148,000 square kilometres of the Ottawa River watershed have been occupied by the Algonquin Anishinaabe peoples for millennia (Lawrence 2012; Morrison 2005; Richardson 1993), since the closing of the last ice age and associated glacial melting led to the emergence of the Ottawa Valley (Russell, Brooks, and Cummings 2011). The Algonquin know the river as Kichi Sibi, meaning "great river." Archaeologists have dated roughly 4,500–4,900 years of Algonquin use around the Gatineau River delta (which forms part of the Ottawa River watershed), with seven meeting areas on both sides of the river (Pilon and Boswell 2015). In the settler imaginary, the Ottawa River watershed has come to represent the "heartland" of the Canadian state, where the three founding peoples — Indigenous, French, and British — first met (Wilcox 2018).

Since their early encounter with Samuel de Champlain in the early 1600s, the Algonquins have experienced over four hundred years of extensive contact with European squatters and settlers in the Ottawa River watershed. In the early 1800s settlement concentrated along the river near the sacred Algonquin site of Akikodjiwan, which includes the powerful waterfall known as Akikpautik to the Algonquins and Chaudière Falls to settlers, as well as surrounding islands. As part of an urbanizing frontier on the periphery of empire, Bytown and Hull (which would become present-day Ottawa and Gatineau) grew around a logging industry and the construction of the Rideau Canal, both of which largely served the interests of the British Empire.

The development of Bytown and Hull facilitated the intensification of resource extraction and settlement, as Algonquin territory was licensed to logging companies and surveyed into townships. Mass incursions by white settlers from the British Isles proceeded throughout the 1800s to the point where Algonquins in Ontario faced displacement in every part of their territories (Lawrence 2012). The Kichi Sibi at Akikodjiwan — representing the spiritual, cultural, and economic heartland of the Algonquin Nation — was appropriated without consent or consideration and came to represent the epicentre of the Canadian settler state. In a few short decades, European settlers built the institutional arrangements of state function and power — Canada's executive, legislative, and judicial branches of government, including the Parliament Buildings — on unceded land.

The settler engulfment of the Algonquin homeland within the Kichi Sibi watershed has been accompanied by the remapping and renaming of much of the territory (Gehl 2014). Beginning with the creation of the provinces of Upper and Lower Canada, now Ontario and Quebec, numerous jurisdictional boundaries were imposed through colonial mapping. The Kichi Sibi was converted into "a waterscape of Algonquin division" (Gehl 2014, 42) that produced "Ontario and Quebec Algonquins" (Lawrence 2012, 52), as well as a source of immense wealth for settler society. The fracturing of the Algonquin homeland along provincial lines was accompanied over time by numerous layers of federal, provincial, regional, and municipal jurisdictions, which "carved out spatial patterns of land use and population control that defy easy mapping" (Pasternak 2017, 21). Settler governments used colonial mapping mechanisms and regimes of private property in attempts to eradicate Algonquin nationhood and rights to the land (Gehl 2014; Lawrence 2012).

The partition of the Algonquin Nation through techniques of settler statecraft was not limited to geospatial configurations and also included linguistic, legal, religious, social, and political ones. While these divisions and experiences are complex, the result today is that there are status and non-status Algonquins (those whom the Canadian government recognizes as "Indians" under the *Indian Act* and those whom it does not), and there are ten federally recognized Algonquin First Nations, nine of which are in Quebec. Consequently, Algonquins have adopted varying and complex strategies in response to the eliminatory mechanisms and domicidal techniques of settler colonialism. The point here is not to emphasize disagreements among the Algonquins produced by the divide-and-conquer techniques of settler colonial governance and property relations, the "many layers of colonial history and policies" (Majaury 2005, 45) that attempt to eliminate Indigenous connections to land that enriches an ever-developing settler society. Instead, I aim to highlight the long-standing and ongoing contestations over unceded and unsurrendered Algonquin territory, especially within Ottawa, and the role of urban development and municipal governance in the reproduction of urban space.

Tools of Dispossession

How exactly does Indigenous land get converted into settler property, and how did that play out in the neighbourhood that came to be known as Heron Gate? Briefly charting the appropriation and development of

Algonquin lands in what would become the southern part of Ottawa helps illuminate two forms of domicide that (1) unmake Indigenous homelands and remake them into settler property and (2) unmake the homes of a racialized migrant community renting shelter on this property.

European invasion and settlement were marked by the creation of a new regime of property relations relying on legal fictions and colonial ideologies that the land was empty and Indigenous people incapable of productively "owning" property. Armed with geographical tools such as the survey, the map (cadastral), and the grid, the British transformed Indigenous land into settler property (Blomley 2003; Harris 2004). Though Indigenous Peoples resisted and disrupted the surveying process, the imposition of a new settler geography — through acts of surveying and mapping Indigenous land — was a "pervasive disciplinary technology" (Harris 2020, 259). A legal (in the eyes of the British) order of property resulted, allowing a system of Indigenous land allotment to settlers (Blomley 2003; Fawcett and Walker 2020).

In what is known in colonial discourse as the "Upper Canada Land Surrenders," the British appropriated land in what has become eastern Ontario. The lands within the current geographical boundaries of the city of Ottawa were never ceded or surrendered by the Algonquins, but the British justified the surveying and sale of the territory through a series of "land surrenders." The "Rideau Purchase" covers some land in the west of the city and was made with a small group of Mississauga people, the details of which are well documented (Crown-Indigenous Relations and Northern Affairs Canada 2016). The "Crawford Purchase," or "gunshot treaty," on the other hand, was made with one Mississauga Elder for a fraction of land north of the St. Lawrence River. The story goes that the Mississauga Elder agreed to share land with the British — as far north from the river that a man could walk in one day, or as far as a gunshot could be heard — in exchange for some clothing and rifles (Reimer 2019; Surtees 1983). While this arrangement pertained to a small fraction of territory, colonial mapping today extends the Crawford Purchase across the entirety of eastern Ontario and north to the Ottawa River. Moreover, these "purchases" were consummated without Algonquin presence or consent (Huitema, Osborne, and Ripmeester 2002). Aside from a letter discussing the arrangement, no other documentation — legal, deed, or otherwise — exists. In other words, British and Canadian claims to this land are dubious at best.

Settlers reconfigured Indigenous land into districts, counties, and townships, which were in turn divided into concessions and then lots that were sold to settlers. This all-encompassing system "constituted the Procrustean bed upon which the new system of land-control and ownership was developed," write Huitema, Osborne, and Ripmeester (2002, 90). "Land became reconstituted as parcels of property, fixed spatially on cadastres, legalized in Land Registries, and legitimated by periodic municipal tax-assessments and government censuses." In the late eighteenth century, Algonquin lands in Ottawa were incorporated into the Gloucester Township. An early survey map of the township, which can be obtained from Ontario's Office of the Surveyor General, shows that the land was parcelled into lots and sold to settlers as early as 1857.

The map also demonstrates the settler colonial approach to property relations through a "schedule of improvements" that documents fee simple ownership of stolen Algonquin land through the practice of surveying. About a century prior to the development of Heron Gate into a rental community, the land was surveyed and sold to Patrick Finn and William Heron. For Blomley (2004, 122), this kind of map plays an "important persuasive role in displacement, both by conceptually emptying a space of its native occupants, and by reassuring viewers of the unproblematic and settled occupation of urban space by a settler society." The implementation of a private property regime on Algonquin land redefined the social, spatial, and legal relations around land ownership. "The normalization of a Western property regime was part and parcel of the spatial imperative of colonization," as Tomiak (2011, 102) puts its. Indigenous dispossession and the introduction and cementing of private property relations were essential to the development of cities as spaces of white settlement and entitlement.

The Reproduction of Settler Colonial Urbanism and Racial Capitalism

The implications of settler urbanization on unceded and contested Indigenous lands are multifaceted. They include the expropriation of Indigenous territory and resources, the establishment and maintenance of hierarchical social relations undermining Indigenous Peoples' self-determining authority (Coulthard 2014), and the removal, or deterritorialization (Tomiak 2011), of Indigenous Peoples from emerging settler urban environments. Combined, these processes amount to the clearing and transformation of land for settler use and "the incorporation of that territory into the regulatory ambit of settler institutions of governance" and accumulation (Hugill 2017, 6). Urban settler formations are premised on the elimination of the original inhabitants and their structures of governance, political orders, spiritual places, and lifeworlds. While a settler colonial politics of recognition (Coulthard 2014) is increasingly normalized through the display of overtures and acknowledgement of Indigenous traditional territories, the replacement of Indigenous political and social orders with a majoritarian settler polity continues.

Settlements are often located on sites of significance to Indigenous Peoples (Blomley 2004), as Ottawa and Gatineau are. Towns and cities played a major role as the "nerve centers" of developing settler states, where the machinery of state institutions was established and where colonial ideas, practices, and commerce were expanded (Gagné and Trépied 2016, 3). Within the colonial imagination, the settler city symbolizes development and progress, liberalism and democracy, and serves as a hub of globalization, international migration, commerce, capital, and power — all of which conceal the city's violent colonial foundations and present (Porter and Yiftachel 2019). Discursive renderings of urban improvement are still prevalent today, as evident in the talk about liveability and revitalization surrounding the Heron Gate redevelopment and the City of Ottawa's New Official Plan.

The creation and reproduction of settler colonial space is not confined to a historical moment or event. It is an ongoing process of dispossession, negotiation, transformation, and resistance, and thus unstable and incomplete (Blomley 2004; Simpson and Bagelman 2018). Through occupation and settlement, settlers attempt to create white places and white spaces (Blomley 2004). Whiteness as a colonizing force is rooted in racial logics of possession, white possessive logics (Moreton-Robinson 2018) that drive the making of countries and cities. As Walcott (2021) notes, McKittrick's (2011, 2013) writing about the plantation—where the duality of white possession converts Indigenous land and Black people into property for settler enrichment—is informative for how contemporary social relations are organized. A growing body of scholarship examines the interconnectivity of race, capital, and space inherent within settler logics of dispossession and displacement in urban formations (Dantzler 2021; McClintock 2018;

Miller 2020; Rucks-Ahidiana 2021). Settler colonial city-making requires continuous strategies of racialization and spatialization, including discursive techniques for depicting Indigenous Peoples as savage and Indigenous spaces as uncivilized (Edmonds 2010). Settlers see Indigenous Peoples and spaces as deviant and out of place, as "not native to the space (or idea) of the settler city" (Cavanagh 2011, 157). For Blomley (2004, 119), "settler cities have also long been imagined as spaces of civilization, set against a world of savagery," with the colonial town imagined as an outpost of progress within the wilderness. Settler colonial city-making has relied upon dichotomous narratives of civilized/savage and progress/primitivism (Mar and Edmonds 2010; Tedesco and Bagelman 2017). Rendering Indigeneity as antithetical to progress and improvement naturalizes dispossession and racializes Indigenous Peoples in colonial territories, where "racial and spatial ideologies coalesced to form part of the bedrock foundations of settler colonialism" (Cavanagh 2011, 154). Ideological discourses continue to be deployed in urban settler formations and real estate development projects that dispossess and displace Indigenous and racialized populations, discourses that work to erase and despatialize Black peoples' sense of place (McKittrick 2006).

Settler colonial spaces are defined and constituted by a racialized, hierarchical relation to land, an interplay between processes of dispossession, accumulation, and white supremacy. In the settler city, these processes render racialized "exogenous Others" (Veracini 2015, 4) as "placeless" within a structure of geographical erasure that eliminates Indigenous Peoples' claims to their homelands. Citing Razack (2002, 49), McClintock (2018, 4) notes that the dominant settler majority must simultaneously (re)create its identity and assert its superiority "against the bodies of racialized Others." The racial and spatial practices involved in developing and reproducing urban formations underlie the complex relationships of settler colonial urbanism (Dorries et al. 2019). This process works to consolidate a particular system of production and social reproduction that is grounded in racialized property relations (Tomiak et al. 2019). Settler colonial urbanism provides a way to think about how racialized settler-colonial violence and dispossession is enacted and reproduced in the making of cities (Dorries, Hugill, and Tomiak 2019).

The social reproduction of property relations in settler societies is facilitated through processes of racialization and accumulation — of racial capitalism — that work to hierarchically configure humans, spaces, and places (Gilmore 2007; Melamed 2015; Robinson 2000). Space, property, and violence are performed simultaneously to dispossess Indigenous Peoples of their land (Blomley 2004) as well as displace undesirable populations from these confiscated lands. Dorries, Hugill, and Tomiak (2019) place racial capitalism and settler colonialism in conversation with each other, elucidating the central role of racialized property relations and real estate development in settler capitalism. The construction and reproduction of racialized hierarchies and legal regimes of private property — as two intersecting and mutually constitutive ideologies — are crucial to the ongoing project of settlement, dispossession, and exclusion (Dorries, Hugill, and Tomiak 2019). Underlying dispossession and displacement in processes of settler colonial urbanism are racialized property relations, "for it is through property that dispossession and settlement are performed and creatively enacted every day" (Blatman-Thomas and Porter 2019, 42). According to Osterweil (2020, 16),

> Racial settler colonialism is thus at the core of all modern notions of property. All our beliefs about the righteousness of property, ownership, and commodity production are built on the history of anti-Black violence and settler-colonial extraction. The right to property is innately, structurally white supremacist: support for white supremacy involves a commitment to property and the commodity form.

By understanding settler colonial urbanism as a process and racial capitalism as the economic-political system in which settler colonial urbanism is enacted, I locate property relations at the heart of settler colonialism's underlying logic (of accessing, claiming, and developing land) and racialized struggles in urban space against housing insecurity. Racialization, as a social process that hierarchically orders the value of life (Story 2019), is deeply implicated in regimes of property relations. Racialization and racial capitalism are central to how settler colonial urbanism reproduces itself through the elimination of racialized spaces that do not conform to settler property ideals and the reconfiguration of those spaces according to their highest and best use. Settler states aim to eliminate racialized spaces through attempts to exclude, isolate, stigmatize, and disperse "non-preferred" (Thobani 2007) migrant populations who are perceived as obstructing urban development, progress, and the perfecting of property relations. Heron Gate is a neighbourhood where these processes are playing out.

Reverberations of Empire: The Coloniality of Migration and Settler States

Heron Gate Village was built as a large, predominantly rental community in the 1960s. Beginning in the 1990s, it developed into a majority-racialized, largely migrant community as families from Africa, the Middle East, and South Asia fled situations of violence and conflict. Ethnic or ethnoracial enclaves arise from a spatial concentration of cultural institutions and businesses in diasporic communities (Mensah 2019). This designation can be also ascribed to communities with a high density of ethnic minorities in a given neighbourhood who share a common heritage, such as culture, language, ethnicity, or custom (Mensah 2019; Walks and Bourne 2006). While ethnoracial enclaves can develop as a result of voluntary or involuntary segregation, contemporary flows of migrants and refugees all trace back to legacies of empire.

In the wake of European imperialism remains a patchwork system of postcolonial and settler colonial states and dominions throughout the globe. Centuries of European imperial conquest resulted in the extraction of vast amounts of wealth from and created enduring divisions in many countries, with violence exacerbated by contemporary and ongoing military ventures into the former colonies. Consequently, Western Europe, along with the North American and Australasian settler colonies, have experienced waves of migration in recent decades. Examining the British state's response to asylum seekers as an example of what Quijano (2000) coined "the coloniality of power," Mayblin (2017) illustrates the continued privileging of white mobility and obstruction of racialized mobility as colonial ideologies informing the treatment of immigrant and refugee populations. The coloniality of power has clear implications for current migration flows, diasporic space, and the establishment of ethnoracial enclaves in settler states.

Critical approaches to diaspora can further our understanding of how migrant populations become subjects of empire and subjected to domicide in the lands where they seek asylum. Brah's (1996) theorization of diaspora enables an investigation into the relationalities of population migrations across fields of social relations, subjectivity, identity, and power. Here, Brah (1996, 16) links the concept of "diaspora space" and entanglements of "genealogies of dispersion" with those of nativist subjectivities and discourses, which in the settler colonies came to be represented by British diasporas (English, Welsh, Scottish, Irish), even though these were internally differentiated by class, gender, and ethnicity. These differences, however, were subsumed under discourses of "Britishness," where British identity was asserted to be superior to Indigenous populations (Brah 1996; Cohen 1996). At the same time, racialized refugees and asylum seekers — despite different origins and backgrounds — can navigate their new homes and neighbourhoods by way of negotiating and reconfiguring of cultural identities based on mutual difference and shared struggle (Hall 1990). Black Canadian diaspora studies are insightful here for understanding how diaspora communities resist being subsumed by the dominant white culture and produce space and belonging in now-white majority countries (Brand 2001; McKittrick 2006; Walcott 2023). In settler colonial states, the cohabiting of space between Indigenous, settler, and migrant populations represents an entanglement of genealogies of dispersion as reverberations of the British Empire (Crosby 2021b).

The reverberations of empire — or relations of power embedded within discourses, institutions, and practices — implicate white settlement and dominance, the attempted elimination of Indigenous Peoples, and the processes of exclusion that mark certain racialized populations as suspicious and undesirable. For Brah (1996, 205), "Diaspora space is the point at which boundaries of inclusion and exclusion, of belonging and otherness, of 'us' and 'them', are contested." Approaching diasporic space as genealogies of dispersion and belonging can help inform relational understandings of the coloniality of power and migration as it pertains to the inscription of racialized modes of subjectivity and identity. It can help us to understand the formation of settler states, contemporary migrancy, the voluntary and involuntary segregation of diasporic communities in ethnoracial enclaves, the role that property relations play in disrupting these enclaves and dispersing racialized populations, and resistance. Together, Indigenous dispossession and racialized exclusion inform our understanding of how racialized migrant populations are coded as nonpreferred others, and how their established communities — such as at Heron Gate — are subjected to domicide.

CHAPTER 3

Domicide in the Liveable City

To understand the complex processes that have led to the targeting of Heron Gate Village for revitalization and gentrification, this book weaves together the concepts of liveability and domicide with understandings of racialized property relations under settler colonialism. Liveability, as emphasized in the City of Ottawa's (2021b, 2) bold entrepreneurial plan to remake Ottawa into "the most liveable mid-sized city in North America," forms the political and discursive urban context in which Heron Gate is destroyed and remade, and its inhabitants displaced and replaced. Working with Porteous and Smith's (2001) concept of domicide as the destruction of home, in this chapter I reconceptualize domicide as a process that works alongside the idea of urban liveability, where urban redevelopment and revitalization projects work to unmake homes, communities, and homelands for some (marginalized, racialized, and Indigenous populations), and remake homes, communities, and homelands for others (affluent, white, settler populations). Racialized property relations are the impetus behind revitalization efforts. Urban improvement discourses are mobilized to produce life, flourishment, enjoyment, and fulfillment for some, and conditions of death, marginality, displacement, and disposability for others. Alongside the idea of liveability are corresponding representations of dilapidation, ghettoization, and criminality, which underwrite the domicidal forces that displace the communities of the racialized and the poor.

Urban Liveability

The concept of liveability has been deployed and harnessed by a variety of political and economic actors, as well as social movements, for decades (Alexander 1970; Cassidy 1980; Jacobs 1961). Urban liveability is broadly defined as the extent to which the physical and psychological needs and demands of a city's residents are met (Chiu 2019). Thus, how people define or attempt to improve liveability is subjective (Kaal 2011; Pacione 1990). For Ley (1990, 34), contestations over what liveability means reveal "much about the various publics who have competed for the power to define the quality of urban life." Liveability could mean entirely different things to different urban residents depending on their income levels. According to Hankins and Powers (2009, 848), the idea of urban liveability "suggests that there is an ideal relationship between the urban environment and the social life it sustains," where the concept of liveability has often and historically been used interchangeably with the notion of "quality of life."

Liveability as a concept has seen mounting popularity in public policy, urban planning, and academic circles in recent decades. Over the last twenty years, the notion of urban liveability has been used as a performance indicator in service of business, marketing, and competition (all components of entrepreneurial governance), with less emphasis on the daily lives of urban dwellers. All scales of government internationally are paying attention to liveability, as cities are more frequently ranked according to various measurements of wellbeing and quality of life (Leach et al. 2016; Ooi and Yuen 2009). Liveability rankings — such as the Economist Intelligence Unit's Liveability Ranking, the Mercer Quality of Life Index, and the more recent Global Liveable and Smart Cities Index — have been criticized for emphasizing business, marketing, and investment indicators that say little about residents' experiences or quality of living (Conger 2015). These types of quantitative indexes assume that human wellbeing or satisfaction involve a certain level of affluence, and may overlook or ignore systemic social and racial inequalities. They fail to account for gentrification, eviction, and displacement, even as officials use liveability rankings to market their cities.

When officials deploy the concept of liveability, profit is prioritized over people. This has led to the observation that liveability is often invoked in city planning and municipal discourse "to describe every urban nicety except the two most closely aligned with people's ability to live — the prices of labor and shelter" (Stein 2019, 69). Critical research has shed light on how liveability is used by officials to justify entrepreneurial policy initiatives (McCann 2004), and by business improvement districts to support their revitalization strategies (Ward 2007). Critical studies have attended to the role of liveability discourses in social production, urban politics, strategies of governance, and gentrification (McCann 2007; McGuirk and Dowling 2011; Tolfo and Doucet 2022). How can we think about ideological discourses of urban improvement — such as liveability and revitalization — in relation to settler colonialism and other critical issues related to race, space, and property? Judith Butler's thinking about liveability and precarity offer a starting point.

Which lives are deemed liveable, and which can be expunged in particular sociopolitical contexts? For Butler (2004), precarity and liveability are distributed unequally in contemporary social and political relations and institutions. Precarity is a politically induced condition in which "certain populations suffer from failing social and economic networks of support and become differentially exposed to injury, violence, and death" (Butler 2009, 25). When social structures or institutions fail or are withdrawn, conditions of unliveable precarity ensue (Butler 2020). Precarity and precarious life form the basis for Butler's (2020) thinking around liveability, which is assigned or attributed to an individual or group of people, or not, either by a group of people or within the terms of a discourse, a policy, or institution. McNeilly (2016) describes Butler's idea of liveability as "the ability to sustain a viable social life in conditions of inherent precariousness and the socio-political operation of precarity." A viable social life — a liveable life — is first determined in terms of access to the basic necessities of life (e.g., food and shelter) but also by "conditions which shape who may be recognised within contingent socio-political cultures as a subject capable of living a life that counts" (McNeilly 2016). A life can only be rendered liveable when certain conditions are fulfilled.

Liveability, when used as an ideological discourse of urban improvement, is creating conditions where life can persist and flourish, but in actuality liveability is distributed unevenly based on an individual's social position, including their racial, class, and gender identity. Applying notions of liveability to both property and people — by planning for their improvement — mobilizes gentrification practices while generating a racial ordering of life. The planning, policies, and processes that go into creating a liveable city involve deciding who gets to live where and whether they qualify for liveability or disposability. In the liveable city, certain people are deemed worthy of flourishing and experiencing a liveable life, while others can be excluded from their homelands or expunged from a community.

Domicide and the Unmaking of Home

This book carves out space in which to explore the production of the liveable city — or urban liveability — in terms of whiteness and property, or what Lisa Guenther (2019) has called "whitespace." Drawing on the work of Elijah Anderson (2015), who examines how white spaces are produced, Guenther describes whitespace as both enveloping and expansive. Jokic (2020, 15) argues that settler whitespace has a third element: "a claim of belonging of the white settler to the territory," where a white sense of entitlement to property serves to racialize space. The production and reproduction of whitespace in settler societies includes the theft of Indigenous land and its conversion into private property, as well as its enclosure, defence, and expansion. We can also conceive of settler whitespace as including a drive to enhance wealth accumulation through the improvement of property (desirable property tending to be in proximity to urban centres), the removal of undesirable subjects who reside on this property as renters, and their replacement with more ideal subjects. Property is central to the production and reproduction of whitespace under settler colonialism and racial capitalism.

Razack's (1998, 2000) theorization of settler subjectivity, space, and violence is informative for thinking about how discourses of urban improvement work to reproduce both space and subjects. Razack's examples of the spatiality of producing, or emplacing, settler subjectivity relates to the physical violence that white settlers inflict on Indigenous Peoples. Thielen-Wilson (2018, 498–99) extends this understanding to incorporate the racializing force of property as integral to the production of settler subjectivity, where "space shapes (re-makes) settler subjectivity as emplaced" and "settler subjects also shape (re-make) space through their violent practices." Settler-emplaced subjectivity is produced through the remaking of urban space, which in turn shapes Other-racialized subjectivity as displaced. Whitespace and settler-emplaced subjectivity are critical to revitalizing and reconfiguring urban settler formations into contemporary visions of the liveable city.

The displacement of racialized subjects and their replacement with settler subjects necessitates the destruction — or unmaking — of their homes and communities, otherwise known as domicide (Porteous and Smith 2001). Domicide denotes the unmaking of home and is typically characterized by repressive elements. Porteous and Smith (2001, 12) define domicide as "the deliberate destruction of home by human agency in pursuit of specified goals, which causes suffering to the victims." In their flagship book, Porteous and Smith (2001, 20) examine over two hundred cases of domicide in seventy countries, ranging from a single dwelling to entire ethnic homelands, in order "to ensure the emergence of an adequate global picture of the motives for, processes of, and results of domicide." For Porteous and Smith (2001, 15), "domicide has a special trauma, because the victims are not killed but must watch their homes being destroyed as they are wrenched forcibly from them followed by an attempt to overcome relocation trauma and to build a new life." Domicide denotes the "murder" or the "deliberate killing of home," but its nature is not necessarily to kill or to end life. It remains distinct from topocide — the murder of place — and genocide — the deliberate killing of a group of people.

Domicide's central concept of home (*domus* in Latin) has been studied widely across disciplines and is a central theme of human geography. Prior to the 1960s, social scientists focused on the physical manifestation of home (the dwelling) yet recognized the "greater depth of meaning" of home (Porteous and Smith 2001). Drawing on the work of Rybczynski (1987) and others, Porteous and Smith (2001) examine the concept of home across a variety of scales to understand home as a space and place constituting centredness and identity for both individuals and groups. This multiscalar typology includes home as place, home as symbol, the psychological meaning of home, and the relation of home to domicide.

Domicide as a sociospatial process of home unmaking (Baxter and Brickell 2014; Nowicki 2017) is a useful concept for exploring ongoing settler colonial expansion and nation-building in relation to policies, processes, and institutions facilitating eviction, displacement, and dispossession. This includes approaching property relations and urban development as driven by settler colonial logics of acquiring Indigenous lands for the exclusive benefit of settler society, and understanding the role of domicide in the reproduction of racialized space. The two largest residential developments in Ottawa at the time of writing, Heron Gate and Zibi, are illustrative here. The Zibi development, for example, involves the federal transfer of sacred Algonquin land (Akikodjiwan) to a private developer despite decades of Algonquin demands to have the land returned (Crosby 2021a). Branded as "world-class," "master-planned," and a "waterfront city," Zibi is being marketed as the most sustainable community in Canada and one of the most sustainable communities in the world. The National Capital Commission touts the development of this "national interest land mass" as key to the revitalization of Ottawa's downtown core. The development will have two thousand high-end condominium units, one million square feet of commercial space, and three million square feet of density. When completed, it will facilitate a population shift of some five thousand settlers — or "pioneers," as the developers have called them (Osman 2018) — to take up residence on the site. The Zibi development is a clear example of settler colonial violence and Indigenous dispossession, an attempt to produce settler whitespace on the spiritual home of the Algonquin Nation.

Domicide is also a useful concept to explore how the homes of racialized, diasporic communities are unmade. What is happening in Heron Gate is the physical destruction of a working-class neighbourhood as well as the symbolic destruction of diasporic space. The demolition of Heron Gate disperses already marginalized populations to the margins of a sprawling city while also unmaking strong sociocultural networks and institutions that have come to define the neighbourhood. Many of the Heron Gate homes demolished in 2018 were led by single mothers, and women may be most adversely impacted by the dissolution of community bonds and networks of support and care, especially in the areas of childcare, enterprise, and reliance on shared first languages. Domicidal forces are remaking tattered townhomes into twinkling towers and replacing Black and Brown communities with a whiter, more affluent demographic.

The Reproductive Side of Domicide

Domicide is about improving urban spaces and populations, and reproducing white spaces and white subjects. It is often associated with explicit violence and destruction (Akesson and Basso 2022) but not often with narratives justifying urban redevelopment efforts — narratives structured by progress and improvement. While the unmaking of Heron Gate is an undeniably violent act characterized by mass evictions and expulsion, it also includes a reproductive aspect where developers remake the neighbourhood into a "liveable" zone of affluence through reproducing space, improving property relations, and replacing subjects.

Domicide is first enacted with the appropriation of Indigenous land for resource development and settler urbanization. Settler colonial property relations involve constantly striving to improve space and perfect real estate to maximize the exchange value of land. Settler colonial urbanism involves converting and reconfiguring property for the highest and best settler use. Herein lies settler colonialism's domicidal violence; it targets Indigenous populations, their land, as well as those subjects — typically racialized, immigrant or refugee, and lower-income populations — who are not the idealized Western subject. Settler property relations relegate Indigenous, lower-income, and racialized populations to the margins of housing tenureship — into rental housing that is likely to be inadequate and unsuitable — and subject them to gentrification and demolition. Within the hierarchy of housing tenureship, where homeownership is the pinnacle of settler success, renters, especially those who are nonpreferred racialized migrants, are subjected to precarity and precarious life.

Improvement discourses, such as those about the liveable city, perpetuate housing precarity by ignoring or downplaying its root causes. That is in part because liveability, as an ideological discourse of settler colonial urbanism and racial capitalism, is primarily concerned with settler vitality. Although masked within the language of improvement in Ottawa's New Official Plan, such as the notion of "injecting new life" into certain urban areas (City of Ottawa 2019b), liveability is about improving property for profit and managing populations. "Injecting new life" involves the revitalization of Indigenous lands and migrant communities for settler prosperity and enjoyment. The rhetoric around urban improvement and the New Official Plan — such as liveability, intensification, renewal, revitalization, resilience, regeneration, and sustainability — does not acknowledge the living situations of lower-income people, does not prioritize renter households, does not offer protection to below-market rental units being lost at record paces to demolition and conversion, and does not seriously engage with options to build new affordable dwellings anywhere close to the numbers needed. In this vein, liveability initiatives can reproduce urban marginality and subject lower-income, racialized, Indigenous, and migrant populations to gentrification and domicide.

The liveable city is an emergent vision for settler society, where the self-perpetuating project of settler colonialism seeks to perfect both Indigenous dispossession and the improvement of property for particular settler subjectivities. This framework is useful to analyze urban revitalization efforts — efforts aimed at producing and reproducing the vitality of settler society through the improvement and perceived best use of property — and applications of domicide in the liveable city. In particular, which subjects are afforded access to quality housing and which subjects are excluded? Which subjects are rendered so disposable that they can be expunged from their homes and their communities in acts of domicide? In other words, whose lives count as liveable in settler society, and whose lives do not?

Although hundreds of lives have been deemed disposable and displaced from Heron Gate to allow for redevelopment, officials do not recognize revitalization efforts as acts of forced dispossession or "violent expulsion" (Sassen 2014). Instead, the redevelopment of Heron Gate has been encouraged by public officials, facilitated by mechanisms of municipal governance, and ultimately propelled by urban planners. Although redevelopment projects are initiated by private developers, the municipality (and larger settler society) establishes the conditions and lays the groundwork for gentrification and domicide. I refer to this symbiotic property relationship as a municipal-developer nexus that, in the service of racial capitalism and settler urbanization, works to determine who gets to live where and under what conditions. The capture of rental housing by financialized real estate and the City of Ottawa's efforts to create a liveable city are deeply shaped by a drive to create whitespace through processes of gentrification and domicide.

Gentrification and Settler Vitality

Gentrification processes are driven by social, cultural, political, and economic forces that shape who gets to live where, who gets excluded and included, and who is able to navigate housing markets and the built environment with more ease and privilege than others. Much has been written on the impacts of gentrification on Black communities, including in the areas of socio-economic, cultural, and health (Freeman 2006; Hyra et al. 2019). Debates about gentrification span decades and have been animated by the key issue of displacement (Brown-Saracino 2013; Freeman and Braconi 2004; Smith and Williams 2013). Critical scholars have attempted to address the argument — put forward in particular by certain urban planners, planning scholars, conservative think tanks and politicians, and economists (see Slater 2009; Stein 2019) — that displacement is a negligible feature of gentrification (Newman and Wyly 2006; Slater 2011). Henig (1980, 638) describes gentrification-induced displacement as "the process in which the poor, minorities, the elderly and the moderate-income working class may gradually be squeezed out of their neighbourhoods." For Stein (2019, 43), gentrification involves three sets of actors: the state as planners of gentrification through revitalization initiatives and land use policies, developers as producers of gentrification through the manipulation of land and the built environment, and the affluent — typically higher-income, white demographics — as consumers of gentrification.

Gentrification, like domicide, works to unmake homes for some while remaking those spaces as homes for others. Gentrification is a process of "un-homing," argue Elliott-Cooper, Hubbard, and Lees (2020), as it always ruptures the connections between people and place. Gentrification and domicide are facilitated by discourses of improvement that invoke life and vitality, such as liveability and revitalization. This process was initially identified in the late 1970s and 1980s (Auger 1979; Palen and London 1984), suggesting that revitalization as gentrification is not a new emergence. At the same time, with the current push to revitalize neighbourhoods in Ottawa and other cities, new discussions have emerged around the myth of "revitalized" urban spaces (Burns and Berbary 2021), the idea of "just revitalization," and the exacerbation of social and spatial inequalities (Ehrenfeucht and Nelson 2020). The revitalization of urban space, as the Heron Gate case illustrates, is about reproducing the vitality of settler society and its most sanctimonious element — property.

Yet settler colonialism, like domicide and gentrification, also embodies repressive and reproductive elements. While settler colonialism relies on the coercive authority of the state to dispossess Indigenous Peoples of their land, it also has a reproductive character that works to produce what Coulthard (2014. 152) calls "forms of life" that naturalize settler colonialism's constitutive hierarchies. Settler colonialism attempts to produce and reproduce land and life — land in the form of property and real estate, and life in the form of ideal white subjectivities. Through property relations, settler colonial urbanism works to produce and improve social spaces and places for white consumption, while nonideal racialized and migrant populations can be subjected to domicide.

The roles of land and life (of producing and reproducing property and white or racialized subjectivities) are central to the social, political, and economic relations of settler colonialism. For Ellis-Young (2022, 2), "gentrification is not just a contested class-based remaking of urban space that displaces lower-income residents, but also a contested remaking of urban space that perpetuates the dispossession of Indigenous people from their land." She argues that gentrification research must be attentive to settler colonial relations so as not to reproduce Indigenous erasure and settler dominance over urban futures, both in scholarship as well as at the local level of resistance. However, some scholars have contested the notion that gentrification is a form of colonization, challenging gentrification researchers to move beyond mere abstractions, metaphors, and false equivalencies (Ellis-Young 2022; Launius and Boyce 2021; Quizar 2019). Instead, they have approached gentrification in terms of its racial and colonial dimensions (Kent-Stoll 2020), and racialized property relations in terms of whiteness, which Montalva Barba (2021) argues can help illuminate the role of white supremacy in gentrification processes. Centring the heightened value of whiteness and the devaluation of nonwhite lives (Launius and Boyce 2021) in thinking around gentrification is useful for unpacking how settler colonial property relations are reproduced in urban redevelopment and revitalization projects. Gentrification entails the social reproduction of settler colonial relations to land (Parish 2020) as real estate.

Processes of revitalization are animated by white racial logics that strive to enhance settler vitality through the creation of liveable urban spaces (e.g., whitespace) and liveable, white subjectivities. Revitalization efforts at Heron Gate demonstrate the enactment of domicide through a discourse of improvement. Revitalization is about producing new forms of life (liveable subjects) and dispersing disposable, nonpreferred subjects. Revitalization is also about the vitality of a given place, and suggests that Heron Gate once had life but is now an unviable (unprofitable) and unliveable neighbourhood. The discourse of revitalization — of improvement, vitality, and the creation of "liveable homes" (City of Ottawa 2018) — obscures the violence and destruction it entails.

Liveability is another ideological discourse of improvement that works alongside domicide to reproduce urban space and emplace settler subjectivities, to unmake the homes and social spaces of those racialized people who do not qualify for wellbeing and are thus determined to be disposable. Domicide is an intended outcome of strategies to produce and reproduce liveable spaces and subjects. As such, domicide is perpetrated by a nexus of public and private actors, including municipal officials and property developers and landlords. The following chapters intricately detail how domicide can be, and is, resisted.

CHAPTER 4

Research Methods and Design

This chapter offers a blueprint for doing engaged research with social movements. The methodological approach is inspired and driven by political activist ethnography and movement-relevant theory. Political activist ethnography is a qualitative research method that focuses on how everyday experiences are represented, organized, and mediated through institutional and ruling relations. Movement-relevant theory centres the knowledge produced by social movements and supplements political activist ethnography through research that is relevant and useful for social movement struggles. Both approaches have their roots in institutional ethnography, a method of sociological inquiry that investigates social organization and social relations based on how people's everyday experiences are mediated through texts and discourses.

It's worth spending time exploring the nuts and bolts of data collection and analysis in this research. By employing political activist ethnography, which is part of a larger political project to disrupt ruling relations, my work centres the perspectives of tenants and community organizers with the goal of producing insightful findings useful for their struggles. Though flexible depending on the researcher and the project, the framework of political activist ethnography tends to involve a threefold research design: (1) start with activist experiences (through participant observation and interviews); (2) analyze interviews and field notes to discover references to institutions and ruling relations; (3) trace these ruling relations to uncover their textual mediation (Rodimon 2018). In this chapter, I unpack my approach to participant recruitment and interviews, participant observation and fieldwork, and textual analysis of my large repertoire of archival materials.

Institutional Ethnography

Developed by Canadian sociologist Dorothy Smith (1987, 1990, 2001, 2005), institutional ethnography has evolved into a critical yet increasingly popular method of social research. Against the prevailing sociological discourse of her time, Smith set out to develop an alternative sociology based on her lived experiences as a feminist sociologist active in the women's movement in the 1970s. Institutional ethnography focuses on the social relations that structure people's lives and is empirically rooted in the experiences of people. Interested in how things work and are actually put together as opposed to what happens or why things happen (Smith 1987), institutional ethnography investigates how texts and discourses coordinate and organize the activities of individuals across space and time (Devault 2006; Devault and McCoy 2001; Smith and Turner 2014). Institutional ethnography developed alongside feminist standpoint theory, of which Smith was also an architect, a framework that centres the social situation of experience, knowledge, and power relations, or what Smith coined "ruling relations."

For Dorothy Smith (1990, 2), ruling relations are intrinsic to and manifest from "institutions of administration, management, and professional authority, and of intellectual and cultural discourses, which organize, regulate, lead and direct, contemporary capitalist societies." These diverse and prolific societal apparatuses permeate liberal democratic and settler colonial societies such as Canada. Ruling relations stem from the visible and invisible systems of governance, bureaucracy, and authority. Through her work developing sociological lines of inquiry from the standpoint of the experiencing embodied subject, of women and the women's movement, Smith not only articulated the role of ruling relations but also pioneered the methodological approach of institutional ethnography. For Mykhalovskiy (2018, 297), "institutional ethnographic inquiry begins with and returns to the embodied site of actual people as they experience the world." The embodied experience of people — including how their lives are organized through their encounters with institutions — is central to apprehending social organization and ruling relations. Through such engagement, institutional ethnographers are better positioned to illuminate established and often taken-for-granted practices of governance and rule.

Political Activist Ethnography

While similar to institutional ethnography in that it investigates social relations of rule through textual and discursive institutional mediations, political activist ethnography begins from an activist standpoint. It emphasizes that activists hold particular social and political insights into the institutions and social organization of issues they struggle against. Whereas institutional ethnography seeks to explicate social relations, political activist ethnography seeks to change them by centring the perspectives of activists whose work investigates and attempts to disrupt the organizing logics of ruling relations (Rodimon 2018).

First conceptualized by George Smith (1990), political activist ethnography aims to produce knowledge that is useful for activists and social movements. Drawing from institutional ethnography, the political activist ethnographer's embeddedness in social struggle enables what Smith refers to as a form of sociology that can assist the activism and organizing efforts of the movement in which the researcher is engaged. The political activist ethnographer works to centre activist/organizer knowledge and investigate pertinent sociopolitical questions.

Aligning with political activist ethnography, the activist tactic of direct action moves beyond forms of state-sanctioned dissent to put pressure on ruling relations. Gary Kinsman (2023) has taken up the idea of direct action as a form of activist research through involvement with antipoverty organizing in Sudbury, Ontario. For Kinsman, direct action is not only a form of political activism but also a form of research that is driven by the needs of the struggle. It can contribute to knowledge production through its disruption of ruling relations, thus informing social movement struggles. By working with the Sudbury Coalition Against Property, Kinsman was able to map social relations of struggle and identify various actors, weak points, and contradictions through which to target ruling relations. This work shows the potential of direct action as activist research and the role of knowledge production.

Dynamics of institutional relations of power can be examined in moments of confrontation to explore disjunctures or lines of fault "between our experiences and knowledges of the social world and the ways in which the world is represented in authoritative discourses" (Rodimon 2018, 38). In George Smith's (1990, 631) work, a line of fault is identified between the knowledge and everyday experiences of gay men and the objective bureaucratic domain of a political-administrative regime. Rodimon (2018) also identifies lines of fault in abortion access in New Brunswick, including the disjuncture between federal and provincial health regulations and the provincial government's representation of abortion access and the reality of abortion care. Through the course of my research, I have identified a number of lines of fault, among them the incompatibility between the discursive application of liveability and the reality of the domicidal practices of urban development in Ottawa, as well as the disjuncture between tenant articulations of needs and demands and how they are interpreted, obfuscated, and ultimately ignored by the dominant private (landlord) and political (City of Ottawa) actors despite claims of active listening and consultation.

Movement-Relevant Theory and Activist Scholarship

Social movement participants have long generated research and theory about their struggles and tactics against dominant ideologies and power structures, as well as produced visions for social change. Rodimon (2018) works to connect movement-relevant theory and knowledge production with political activist ethnography using Bevington and Dixon's (2005) work, which underlines how researchers can learn from the epistemological and theoretical work in which activists are already engaged. Knowledge production is an integral aspect of the everyday work and experiences of activists and social movements (Choudry 2013, 2014, 2015, 2019; Dixon 2014; Kinsman 2006). Social movement settings act as key sites of knowledge production where researchers can investigate experiences and practices that are organized by and through ruling relations (Rodimon 2018, 37). Ruling relations can be further examined by analyzing the institutions confronted by activists in the struggle for social transformation.

Political activist ethnography and movement-relevant theory necessarily involve what Bevington and Dixon (2005) theorize as "direct engagement" with movements struggling for social change. Direct engagement involves "putting the thoughts and concerns of the movement participants at the center of the research agenda and showing a commitment to producing accurate and potentially useful information about the issues that are important to these activists" (Bevington and Dixon 2005, 200). Relationship building with movements in the service of producing movement-relevant scholarship necessitates dynamic and reciprocal engagement. Direct engagement enables researchers to develop a deeper and more nuanced understanding of social movements, which is more likely to produce better research and useful theory.

Notwithstanding the potentiality of political activist ethnography as a research method that works to explicate, challenge, and transform ruling relations, there are potential limitations and cautions worth exploring. One challenge is the role of the "activist scholar" (Bisaillon 2012; Mayer 2020) in building and maintaining relationships with activist groups while remaining attuned to the dynamics within and between groups and the importance of maintaining a receptive yet critical stance toward activist ideas and strategies (Rodimon 2018). Conflicting dynamics can emerge within particular activist groups and especially between different activist formations within a particular milieu. Neighbourhood residents and tenants organizing within their communities are understandably suspicious and can even be hostile toward outside organizations that come onto the scene. These organizations may be seen as self-interested, working to divert resources, and distracting from the main goals, for example, by trying to negotiate concessions or a more acceptable form of domicide with ruling actors rather than taking direct action to defend homes and communities. These dynamics were certainly at play in the Heron Gate struggle. Although my point of entry and allegiances lie with the Herongate Tenant Coalition, I engaged with a broad range of relevant actors for this research, to gain a fuller perspective of the struggle and how the numerous actors articulate their goals and operationalize their strategies. The challenge for activist scholars in these types of circumstances is to remain reflexive about our social location and privilege, and how that may impact knowledge production.

My approach to political activist ethnography in this project is one of doing, not studying. I work sometimes from within and often alongside the Herongate Tenant Coalition to confront the landlord and defend the neighbourhood; I did not study the group from the outside. However, I have maintained a formal distance through my independent media work so as to be able to write publicly about the struggle while conforming to established media norms about objectivity and bias; this caveat must be contextualized by my decade-long work with *The Leveller*, a progressive newspaper that publicly confronts and challenges mainstream media claims to objectivity and non-bias. Further, although I have connections with the Heron Gate and Heatherington neighbourhoods, I am not a resident of this area.

My approach in this book is to centre the knowledge produced by tenant organizing as well as the theoretical permutations generated through this work. Much of what I set out to do is to understand the dominant interests at play in this struggle, the ruling relations produced by financial real estate capital and municipal governance actors. In addition, I bring various theoretical lenses to bear on how the landlord-developer manoeuvres and operates within the built environment and the property relations produced therein. Bridging movement-relevant theory with political activist ethnography and investigating social relations of struggle through a diversity of activist tactics help in understanding and articulating ruling relations and also relations of resistance.

Data Collection and Analysis

My data collection and analysis processes were inspired and driven by the research design strategy adopted by institutional and political activist ethnographers, where those conducting movement-relevant, activist research are already engaged in activism and social movement support work (see Kinsman 2023; Rodimon 2018; Withers 2019). This activist-research ethnographic work is naturally iterative and inductive, in the traditional qualitative sense that it is a process of discovery (O'Reilly 2011) but also in the sense that activists are already *doing* the work as opposed to studying the work. Movement-relevant theory becomes useful here to understand activist work as research and knowledge production; the role of political activist ethnographers approaching the work academically is to aid in mobilizing knowledge for the benefit of social struggle against ruling relations.

Political activist ethnography's methodological approach to data collection and analysis is flexible and open (Rodimon 2018). There are three choice methods — interviews, participant observation, and textual/archival research — but proponents of political activist ethnography have argued for the option to use others as well rather than impose strict parameters. This ethnographic approach to doing research is "explicitly political, emergent in design, and embedded in social movement practice" (Mykhalovskiy 2018, 302). Likewise, data collection in both the institutional and political activist ethnographic traditions is considered nonlinear, in that the process of discovery through these methods can lead to other points of inquiry and analysis.

Participant observation

Much of my work with the Herongate Tenant Coalition related to research and fundraising, including submitting freedom of information (FOI) requests to the City of Ottawa and writing news stories for independent media. I first became involved with the Herongate Tenant Coalition in June 2018, shortly after the group formed. Timbercreek Asset Management had announced the demolition of a large parcel of townhomes in Heron Gate Village and issued eviction notices. Soon thereafter, Timbercreek sent a cease and desist letter to members of the Herongate Tenant Coalition and I wrote my first news article covering this story, which was published by national outlet Ricochet Media on July 23, 2021, and then republished by *The Leveller*, a community newspaper based in Ottawa (Crosby 2018a, 2018b). The Herongate Tenant Coalition's small research team then worked to uncover financial and political links between the developers and city officials and planners, filing a series of FOI requests that also investigated City of Ottawa bylaw enforcement in relation to the systemic neglect and lack of maintenance enforcement in the Heron Gate community. This initial work would eventually bear fruit: several years later, in 2021, the head of the city committee in charge of urban planning and development, Jan Harder, would resign her seat as a result of our research and media work, which was followed up on by the city's integrity commissioner (Lord 2021). An FOI request I filed would also become the focal point of a lawsuit brought by former tenants seeking damages from Timbercreek for neglecting to make necessary repairs and then lying about making those repairs to the Property Standards and License Appeals Committee (PSLAC).

In early 2019 I began to keep an ethnographic timeline of events tracking my encounters related to Heron Gate, as well as a notebook for the purpose of crafting field notes. I recorded notable occurrences from the previous year, and gathered and organized associated documentation. To broaden my knowledge and experience beyond one particular microgeographical context, I also immersed myself in the wider Ottawa and Canadian milieus of housing justice activism, advocacy, and education. I joined various networks and participated in dozens of events organized by a range of national groups, including the Centre for Equality Rights in Accommodation (now the Canadian Centre for Housing Rights), the National Right to Housing Network, the Planners Network, and the Expert Community on Housing. I also attended the Healthy Cities forums organized by Carleton University. At a more local level, I became involved with activist group Horizon Ottawa, which advocates for progressive change municipally, and I participated in events and actions organized by grassroots groups such as ACORN Ottawa. At these events I paid particular attention to activists' articulations of what is required to realize a just city where residents have access to affordable housing and are not living precariously under the threat of eviction. I also participated in other tenant struggles, attending protests and carrying out research for groups lobbying for reforms and fighting landlords and the Landlord and Tenant Board in Ontario.

At the time of writing, I had recorded my participation in ninety unique events and documented my field observation work. I organized these events into particular categories, including community forums organized by developers or opponents, strategy meetings organized by tenant activists, Landlord and Tenant Board hearings, protests, webinars, real estate investment conferences, and legal proceedings. Out of these ninety events, I pinpointed thirty-five to take forward to the coding and analysis stage, after dividing them further into coherent categories. These categories include instances related to court proceedings, public "consultation" sessions organized by the landlord-developer, events and meetings organized by the Herongate Tenant Coalition and other community groups, real estate investment conferences, public events organized by housing activists, and City of Ottawa–related events such as council and planning committee meetings and public education or consultation sessions related to Heron Gate and the New Official Plan.

One interesting point of intervention involving my field work in this project has been my participation in court proceedings. I have been involved with organizing, strategizing, and contributing to work on the legal cases emanating from the 2018 Heron Gate demoviction. This experience has added an extra layer of observational analysis through which to understand the legal tactics and strategies deployed by landlords and tenants in court.

Since embracing this activist work as a research project, I have participated in every event relevant to the Heron Gate redevelopment and the City of Ottawa's New Official Plan process, allowing me to gain a comprehensive understanding of the social organization of city politics and urban development. I am also embedded in the organizing efforts at Heron Gate and other tenant struggles related to evictions and rent increases. I have participated on various panels and contributed to wider efforts to document financialized landlords in Ontario and the spike in evictions during the COVID-19 pandemic. I have also continued to write media articles on financialized real estate, evictions, and tenant resistance. From these ongoing experiences I have developed a solid understanding of what I believe is useful for housing rights activists in terms of academic research as well as where I can be most useful doing activist work (FOI research, media articles, fundraising, and activist research documenting financialized landlords and evictions).

Interviews

My approach to interviewee recruitment for this project was purposeful, targeted sampling. I kept and regularly updated a list of prospective interviewees based on my ongoing fieldwork and participation in a wide variety of events. The list included many individuals involved in activism on Heron Gate and tenant justice, as well as representatives from the landlord-developer, City of Ottawa (e.g., councillors and their staff, and housing and planning department personnel), community organizations (e.g., nonprofit and social agencies working in proximity to the community), and other experts in affordable housing, financialization, and urban development. I set out to complete twelve to twenty-five interviews that would provide further insight into the tactics and strategies behind development initiatives and evictions defence, as well as how these groupings of social actors conceptualize and envision liveable communities. By recruiting participants through the networks fostered by my fieldwork, I met my upper target of twenty-five interviews. I interviewed seven people from the Herongate Tenant Coalition, six people working for the City of Ottawa, two people working for the landlord-developer, as well as lawyers, civil society people, non–Heron Gate housing activists, and a non–Heron Gate Timbercreek tenant. While over half of participants were white, and the majority of City of Ottawa representatives were white men, the activist subgroup of participants was mostly women of colour.

Not included in this group of interviewees are the people I spoke with as part of my media work on the Heron Gate redevelopment, evictions, and other struggles around the financialization of rental housing and affordability, the majority of whom were tenants and activists. I was also able to gather sufficient insight from participating in multiple forums where developer representatives made presentations (including community "consultation" events, City of Ottawa–organized events, and real estate investment conferences). Overall, I gathered considerable amounts of data from the dominant actors in this struggle.

Textual record collection

Texts are key components of the architecture of ruling relations. They include diverse forms of writing and representation from a variety of institutional sources, such as media, policymakers, legislators, and courts, and from different organizational sectors, including public, private, nongovernmental, and social movement based. As such, collecting texts is a central focus of institutional ethnography and political activist ethnography. Texts from a variety of sources have played a pivotal role in documenting and assisting me to understand ruling relations and resistance in the ongoing struggle at Heron Gate. I have collected hundreds of documents through online searches, on-site research, email requests, FOI requests, participating in events, and organizing with the Herongate Tenant Coalition. These textual materials come from the three main protagonists involved in this struggle: the City of Ottawa, the landlord-developer, and tenant and housing activists and organizers.

The City of Ottawa is interesting for a number of reasons. First, it refused to intervene in the 2016 and 2018 demolition-driven evictions. Second, it is responsible for permitting the rezoning, demolition, and redevelopment of the Heron Gate neighbourhood, as well as enforcing maintenance and repair standards of rental properties. Third, in September 2021 it approved the Official Plan Amendment submitted by the landlord-developer to redevelop the entire neighbourhood, which includes demolishing 559 more homes (in addition to 230 already demolished) in the coming years. A plethora of textual materials have been created for the Official Plan Amendment and redevelopment agenda, including over forty planning proposal documents submitted by the landlord-developer. Intertwined with this agenda is the charge levied by Heron Gate residents and activists that the landlord-developer purposefully neglected the property and failed to maintain adequate living conditions by refusing to make necessary repairs. I have led the FOI research as part of the Herongate Tenant Coalition's efforts since 2018, a major part of which was attempting to understand the role of the city and the bylaw department in enforcing work orders and maintenance in the neighbourhood. As mentioned above, one of my FOI disclosures has led to a lawsuit being filed against the landlord in small claims court.

I have also submitted FOI requests to the City of Ottawa to learn more about planning, development, and the decision to designate Ottawa as North America's most liveable mid-sized city. I supplemented these internal records with the collection of publicly available annual reports, economic development reports, growth management strategy reports, records associated with the Community Safety and Well-Being Plan, which seeks to address local risks to safety and wellbeing in different priority areas including housing, as well as documents associated with the Building Better Revitalized Neighbourhoods program targeting specific low-income communities in Ottawa (including Heatherington) for improvement. I also specifically targeted documentation related to the development of Ottawa's New Official Plan, which is led by the Planning, Infrastructure and Economic Development Department. Altogether I have collected more than 150 unique documents related to the City of Ottawa's overall planning strategy to pursue economic growth through urban development.

Second, working closely with the Herongate Tenant Coalition, I have collected a large number of textual materials produced by the Heron Gate landlord. Timbercreek Asset Management (now Hazelview) is a financialized real estate investment firm comprised of numerous entities and subsidiaries. Part of the research I have undertaken with the Herongate Tenant Coalition is to understand the composition and strategy of this type of ruling actor as it pertains to reconfiguring the built environment and hypercommodifying housing in order to enrich investors. We have collected promotional materials, media articles, and financial documents spanning the company's history dating back twenty years. In addition, in order to better understand this type of actor, I have also researched and collected documents on other large real estate investment players in Canada. A broad survey of materials — including annual reports, brochures, conference presentations, memoranda, presentation slides, prospectuses, and other materials these firms produce for their investors — highlights the characteristics, motivations, and business strategies of apartment investors and their approach to both urban redevelopment and tenant activism.

Third, I have access to the Herongate Tenant Coalition's extensive repertoire of organizing documents and legal materials, allowing me to conduct extensive document-based analysis. Textual materials include landlord communications with tenants about the 2018 mass evictions, assorted media articles, and the coalition's social media campaign. There is also a trove of legal documents and background and supplementary materials stemming from the five court cases initiated by the Herongate Tenant Coalition. These include all of the legal letters and cease and desist notices filed by the landlord and the Herongate Tenant Coalition, all court submissions (including statements of claim, evidence, and books of authorities) associated with the cases, and documents from my FOI research supporting one of the court cases. This array of materials also includes documentation of neighbourhood demographics compiled through a survey of the 2018 demoviction zone in comparison with Statistics Canada and census data. The neighbourhood demographic survey and census analysis outlining income disparities and core housing need in the neighbourhood is a valuable tool that can be used with other indexes that measure liveability.

Data mapping and coding

Following Rodimon's (2018) lead, my approach to data collection and analysis, using the methodological approaches to sociological inquiry outlined above, was a nonlinear, emergent process of generating knowledge for further exploration. This approach best suited my project because my positioning as an activist enabled me to look at the research from the standpoint of tenant movements that challenge ruling relations. This book is ambitious, tackling a mammoth urban development project where the landlord-developer has deployed multipronged strategies to reconfigure the built environment and suppress tenant opposition. The redevelopment is also informed by multipronged strategies of urban governance, including a historic reconfiguration of the City of Ottawa's Official Plan alongside support for domicide at Heron Gate. Likewise, tenants have used multipronged activist strategies and tactics on multiple fronts.

A challenge of this project has been to sift through and organize large amounts of data. I organized the hundreds of files compiled for this research into four master folders pertaining to the City of Ottawa, the landlord-developer and the redevelopment of Heron Gate, the Herongate Tenant Coalition, and research methods. The first three master folders contained all texts related to the research, while the methods folder contained documentation from events, field notes, and interviews. Together, these four folders contained a combined 2,716 files. I then used a data mapping document to organize the research into main themes: (1) City of Ottawa, municipal politics, and urban development; (2) Herongate Tenant Coalition court ethnography; (3) community organizing and events; (4) Timbercreek and Heron Gate. Using NVIVO software I set out to code my qualitative data for further analysis. I employed a political activist ethnographic approach to coding, which traces connections between the institutions and organizations involved in practising and resisting domicidal ruling relations.

Institutional and political activist ethnographic analytical approaches centre people's experiences in an attempt to understand and connect these experiences to ruling relations (Rodimon 2018). From this process I was able to craft an outline of how the empirical information would take shape based on the various themes, actors, and data. I did this while remaining cognizant of the importance of telling a compelling, chronological, and ultimately coherent narrative that centres the Herongate Tenant Coalition and their actions.

CHAPTER 5

Heron Gate, Racial Stigma, and Strategic Neglect

A profound sense of community exists in Heron Gate — the result of what some call an "ethnic enclave" (Yussuf et al. v. Timbercreek 2019) or "ethno-racial enclave" (Mensah and Tucker-Simmons 2021) — across diasporic space shared by East African, Middle Eastern, and South Asian groups. While socially and culturally nurtured, however, the neighbourhood faces structural discrimination and racialized poverty. With an average median income around half of the city-wide average (Statistics Canada 2017), it has one of the highest concentrations of poverty in the region, as well as some of the highest rates of core housing need in Ottawa. Most residents of the neighbourhood rent their homes. Heron Gate has suffered from racial stigma and physical neglect, both of which have contributed to justifications behind demolishing parcels of the neighbourhood. Below I provide a unique window on strategic neglect at Heron Gate through an investigation of a municipal committee where developers can appeal city-issued maintenance orders without the knowledge or participation of tenants.

A Liveable Community?

The story of Heron Gate's origins and contemporary remaking involves state- and developer-led processes of urban renewal and revitalization. Heron Gate was originally constructed to house those displaced from the working-class district of LeBreton Flats, west of Ottawa's downtown core (Hussein and Hawley 2021), when it was demolished in 1965 in the name of slum clearance and urban renewal (Picton 2010). The "territorial stigmatization" of working-class districts like LeBreton Flats contributed to these areas being viewed as "unnatural 'slums', and also as sources of disease and immorality that negatively affected the quality of family life and citizenship" (Picton 2010, 310). Picton (2010, 305) describes a postwar urban planning agenda that centred renewal and social redistribution, which meant targeting low-income areas for demolition and their inhabitants for dispersal; "urban reconstruction" would transform congested areas and "create new, clean, functional cities." While urban renewal efforts had different flavours throughout cities in Europe and North America, Ottawa city planners' version of urban renewal was inspired by the work of French architect and planner Jacques Gréber (Picton 2010).

Despite the key role that he played as an architect in the Vichy regime, Gréber was commissioned to draw up a design for improving Ottawa's aesthetics. By doing so, he instilled urban ideals that arguably continue to inform planning and development in the national capital region. The 1950 Gréber Plan "provided the design details for a restructured central core, an expansive creation of parks and green spaces (including a greenbelt), the decentralization of industry and housing and the development of an extensive network of parkways and expressways" writes Picton (2010, 310). It also "called for the relocation of inner-city residents in blighted and congested neighbourhoods to pristine, clean, low-density suburban environments. The [plan] would beautify and rationalize the national capital and create a more hygienic and morally secure future for the National Capital Region."

The working-class and lower-income residents of LeBreton Flats bore the brunt of beautification. Urban renewal evictions in the 1960s drove them to the city's margins, which included Heron Gate. A city's outlying areas are important for understanding uneven development and exploitation, as working-class and new immigrant populations lived there in the era of urban renewal (Hussein and Hawley 2021). Heron Gate now forms part of an area called Ledbury–Heron Gate–Ridgemont, which was considered suburban before Ottawa amalgamated in 2001, and today it is considered outer urban or an inner suburb. It has become more central than peripheral, with rising property values reflecting its geographic location.

Just as LeBreton Flats residents were pushed to outlying neighbourhoods such as Heron Gate, working-class tenants in Heron Gate are being displaced to the margins where rents are most affordable (albeit decreasingly so) as the neighbourhood undergoes contemporary gentrification processes of renewal and revitalization. For Komakech and Jackson (2016, 415), urban renewal interventions aim to "inject new vitality" through planned adjustments to the built environment. Renewal, like zoning and other land use policies, is designed to create and maintain rational and orderly landscapes (Stein 2019). However, in reality these urban planning tools, Stein (2019, 28) argues, are often deployed "to target one racial group for exclusion or expulsion while clearing the way for another's quality of life." Revitalization is but a new iteration of renewal invoked more and more in urban planning interventions.

At the time of their demolitions, LeBreton Flats and Heron Gate shared some commonalities, such as inadequate housing, poverty, neglect, and stigmatization. But they also offered a strong sense of home, community identity, and sociocultural support to their residents. Despite facing structural discrimination, they had many qualities associated with liveability.

Ottawa Neighbourhood Study: Ledbury–Heron Gate–Ridgemont

Conducted by researchers at the University of Ottawa, the Ottawa Neighbourhood Study provides data drawn from residents' experiences, perceptions, and knowledge of geographical areas, as opposed to municipally determined administrative boundaries. In the study, Heron Gate falls within the Ledbury–Heron Gate–Ridgemont neighbourhood, which includes Heatherington to the south, bordering neighbourhoods to the east, and adjacent neighbourhoods west to Bank Street. For a detailed description of this larger constituent area, see Xia (2020).

In this neighbourhood 75 percent of residents are renters, in contrast to the Ottawa average of 34 percent. Rates of core housing need are far higher than the average numbers in Ottawa. Compared to the rest of Ottawa, residents of Ledbury–Heron Gate–Ridgemont are more likely to live in unaffordable housing that requires major repairs, and are four times more likely to live in unsuitable housing (Ottawa Neighbourhood Study 2021). Further, residents of all age groups are much more likely to live in low-income households, according to the low-income measure, which captures households whose after-tax income falls below 50 percent of the Canadian median, adjusted for household size. Households in Ledbury–Heron Gate–Ridgemont are over three times more likely to meet the low-income measure compared to the Ottawa average, at 41.2 percent compared to 12.6 percent (Ottawa Neighbourhood Study 2021).

Moreover, demographic and immigration numbers reveal that two-thirds of the neighbourhood population are racialized, compared to just over one-quarter of the Ottawa population. Significantly, 27.4 percent of residents are Black and 15.1 percent are of Middle Eastern descent, compared with 6.6 and 4.5 percent overall in Ottawa, respectively. Reflecting the large immigrant and refugee population, residents are over four times more likely to speak neither English nor French compared to the wider city. More than half of the population are first-generation immigrants, which is double the figures for Ottawa. Immigrants are over 3.5 times more likely to settle in Ledbury–Heron Gate–Ridgemont than elsewhere in the city, and refugees are over five times more likely to settle there; in fact, Ledbury–Heron Gate–Ridgemont houses 22.8 percent of the city's refugee population. It is not apparent, however, to what extent the settlement and segregation patterns are voluntary and to what extent social service and settlement organizations steer newcomers into the community (Ottawa Neighbourhood Study 2021).

These statistics are important for understanding processes of gentrification and domicide in Heron Gate, with its high density of relatively poor people of colour and immigrants. Mensah and Tucker-Simmons (2021, 83) argue that "Herongate's socio-economic composition makes it an archetypal example of an urban space that is susceptible to the encroachments of contemporary gentrification." The Herongate Tenant Coalition and its supporters have persistently argued that the landlord targeted the neighbourhood because of these demographics. They contend that the predatory nature of financialized real estate marks lower-income and racialized areas for transformation, in part because newcomers and those who do not speak English or French as a first language are seen as less likely to challenge eviction.

Heron Gate's census profile

Supplementing the Ottawa Neighbourhood Study are Statistics Canada census data reflecting 2016, which allow for a more tightly focused picture of resident demographics and core housing need specific to Heron Gate Village before the Timbercreek evictions began. In fact, the property is large enough to have its own census tract. The following comparisons between the 2016 census profiles for

Heron Gate and the City of Ottawa reveal even starker differences than did the Ottawa Neighbourhood Survey in terms of race, income, and housing.

Visible minorities, as defined by the Canadian government, comprised 70 percent of Heron Gate's population of 4,681, compared to 26 percent of Ottawa's population of 934,243; 30.9 percent of residents were Black compared with Ottawa's 6.6 percent, 15.2 percent were Arab compared with Ottawa's 4.5 percent, and 11.8 percent were South Asian compared with Ottawa's 4.2 percent. In terms of language, 9.5 percent of Heron Gate residents spoke neither English nor French compared to Ottawa's 1.5 percent. Over half of Heron Gate residents (52.3 percent) were immigrants compared to less than one-quarter of Ottawa residents (23.6 percent). Refugees comprised the vast majority of immigrant types at 65.3 percent as measured between 1980 and 2016, compared to 23.8 percent for the whole of Ottawa during the same period. Migrants were most likely to arrive from African and Asian countries, with Somalia, Iraq, Syria, and Nepal representing the top source countries for recent immigrants (2011–16) at 54 percent of the total (Statistics Canada 2017).

In terms of housing, 93.3 percent of the Heron Gate population were renters, compared to 34.3 percent of Ottawa. Neighbourhood households disproportionately experienced core housing need, measured in terms of suitability, disrepair, and affordability. Regarding suitability — which measures whether a dwelling has enough bedrooms for the size and composition of the household — 27.6 percent of Heron Gate households were living in unsuitable conditions, compared to 4.6 percent of Ottawa households. This is further reflected in the fact that over 12 percent of households had more than one person per room, compared to Ottawa's 1.6 percent. Further, the rate of housing requiring major repairs was disproportionate, at 10.2 percent compared to Ottawa's 5.3 percent. Finally, over half of Heron Gate households (51.6 percent) spent 30 percent or more of total household income on shelter costs, compared to less than one-quarter (23.8 percent) throughout Ottawa (this includes owners and renters). Moreover, the median total income of households was less than half the Ottawa average, at $40,594 compared to $85,981 (Statistics Canada 2017). Overall, these figures paint a grim picture of housing inequality, structural racism, and spatial segregation.

The social fabric of Heron Gate

Beginning in the 1990s, the neighbourhood became home for considerable numbers of immigrants and refugees fleeing conflict in Somalia, then Iraq, then Syria. A significant number of people from South Asia, in particular Nepal, also settled in Heron Gate. The neighbourhood developed into what Mensah and Tucker-Simmons (2021) describe as an "ethno-racial enclave," a concept adapted from "ethnic enclave," as Heron Gate was referred to in the application submitted by evicted tenants to the Ontario Human Rights Tribunal (Yussuf et al. v. Timbercreek 2019). Geographically, an ethnic enclave is an urban area with a high concentration of one or more ethnic groups with shared cultural identities. Sociologically, as a result of the clustering of immigrants and refugees in specific geographic areas, an ethnic enclave develops through the formation of migrant networks and the fostering of intersecting relations and forms of capital (social, economic, cultural) (Pullés and Lee 2019).

Surveying the research of immigrant neighbourhoods in Canada, Xia (2020, 12) identifies a prominent theme: the spatial clustering and concentration of immigrants in particular geographic areas is "influenced by a mix of structural factors and individual agency that leads to a variety of outcomes for immigrant groups that is impossible to generalize." Structural factors may prevent immigrants from settling outside of the ethnic or co-ethnic neighbourhood. The high concentrations of racialized people in a few specific areas in Ottawa, such as Heron Gate, results from structural racism, which includes housing discrimination, employment discrimination, and inequality of income distribution (Crawford 2022).

Research has shown that as was the case for the earlier residents displaced from LeBreton Flats, many settle in the neighbourhood out of necessity rather than choice (Xia 2020, 54). Although Heron Gate is one of the poorest neighbourhoods in the city, it contains a vibrant mutual support network of the kind that tend to emerge in neighbourhoods of concentrated poverty with a shared identity (August 2014a). Newcomers have ready access to existing social supports, networks, and amenities, some of the key necessities of life (Xia 2020, 54). For those living within the community, Heron Gate offers many of the constituent components of liveability. One resident described the benefits of the neighbourhood like this:

> If you walk by one area you smell Nepali food, you walk by another area you smell Somali food, you walk by another area you smell Arabic food. You hear a multitude of languages just walking in the neighbourhood. It's a culturally and ethnically diverse area and there's no area like it.... It's so interconnected and actually is what a community is supposed to be, like neighbours depending on their neighbours for things like food, babysitting, transportation. And so I think that is something that gets lost. It's an extremely unique neighbourhood and people are extremely interdependent and interconnected. (Herongate Tenant Coalition 2020)

Heron Gate offers support networks and a refuge for newcomers, immigrants, and refugees, who can survive and even thrive within an enclave of others who speak their first languages.

Lily Xia worked closely with the Herongate Tenant Coalition while she completed a master's degree at the University of Ottawa. Her research documents a deep yet complex sense of belonging among the diverse immigrant communities that comprise Heron Gate. Residents "have built a sense of belonging and community, providing each other support and solidarity in spite of challenges such as (in)accessibility, discrimination, and disinvestment," according to Xia (2020, 2). Heron Gate residents are keenly aware of the structural barriers that they face; however, they refuse to allow these barriers to define or disrupt their sense of belonging and support for their neighbours. Residents' shared sense of belonging is not based only on ethnic belonging but also on mutual difference and shared struggle. Heron Gate residents have built a sense of community around common difference, "where the differences between individuals are acknowledged but not used as a marker for exclusion" (Xia 2020, 148). This sense of community belonging is demonstrative of the value of home in Heron Gate.

Communities like Heron Gate offer some insulation from the wider exclusionary sentiments of Canadian settler society. The Herongate Tenant Coalition (2018a, 7) describes the community as "a phenomenally tight-knit neighbourhood, where doors are always open, where many members of racialized communities don't feel like the 'Other' and don't have to live always braced for being told to 'go back to your country.'" In an interview, one organizer with the coalition spoke of the valuable social fabric that immigrant and refugee communities in the neighbourhood rely upon, and the impacts of mass eviction and displacement:

> Some of them are brand-new to the country and are trying to still get their head around Canadian culture, which is very bizarre depending on what part of the world you come from, and understanding how health care works, access to childcare, education, religion, faith-based stuff — like all kinds of things that people would normally need to figure out, right? You're coming to a strange foreign land of Canada and people were able to make a better go of it, in particular Somalis coming to this neighbourhood because there is like a network of other people that were there, people who spoke the same language, understood each other. There's the classic example of neighbours, two or three homes, people, women would drop their kids off across the street at one of their neighbour's homes to babysit while they went off to work and then they trade off. So like the community was very reliant on each other. It's like there is actually a huge economic and cultural value towards these people being able to live and relying on one another. And so, you know, people, working-class immigrants brand-new to Canada are particularly vulnerable towards that kind of disruption. (Anonymous interview, June 7, 2021)

The interviewee further described the disruption to the social fabric:

> It's a common integration pattern [for migrants] across the country. People are going to look out for, you know, are going to move to a place where there's actual support from other people, possibly even people they knew from home. In the case of Herongate, there were people who knew each other from refugee camps and met in refugee camps, and then were reunited. That's the shit that gets disrupted by these mass evictions as well ... like tearing down any homes and mass evicting anyone is bad. However, disrupting what I've tried to just describe, disrupting the fact that there are people who know each other from refugee camps and were then reunited in this neighbourhood and then look after each other's children, that thing, which is hard to put a monetary value on, was disrupted and taken away from a vulnerable population.

This statement captures some of the social and symbolic intricacies of how a community gets disrupted, how a community of immigrants and refugees that rely upon each other, and enact mutual aid in common difference, gets eliminated. In these examples, Heron Gate offers a semblance of familiarity in a foreign and at times hostile place. This ethnoracial enclave offers a sense of security, even though outsiders fixate on aspects of insecurity that in their view contribute to making the neighbourhood unliveable. Herein lies a tension between perceptions and realities of what makes a neighbourhood unliveable or liveable. Researchers have highlighted the positive attributes of such enclaves, which can contribute to an enhanced quality of life for their inhabitants in terms of maintaining cultural values and practices, strengthening social networks and social cohesion, and supporting economic enterprises and sociocultural institutions (Mensah and Tucker-Simmons 2021; Mensah and Williams 2017). However, the dominant white majority often views ethnoracial enclaves negatively, associating them with poverty and crime (see Walks and Bourne 2006).

Racial Stigma

As Heron Gate has undergone a demographic and socioeconomic shift since the 1990s, stigmatization of the neighbourhood has grown. Although strong social supports and cultural networks are created by residents, engendering a deep sense of community belonging and liveability, the perpetuation of racial stigma associated with the neighbourhood has worked to justify systemic and strategic neglect. In addition, racial stigma has fuelled a narrative that Heron Gate homes are not worth saving and should be demolished; in this narrative driven by public officials, mainstream media, and the landlord, the neighbourhood is seemingly devoid of liveability.

Apartment Watch and perceptions of crime

Heron Gate has been the focus of crime prevention projects going back to the late 1980s (Meredith and Paquette 1992). One such program that ran in Heron Gate's five apartment buildings from the mid- to late 1980s was Apartment Watch, based on the Neighbourhood Watch model of social control that makes citizens responsible for crime prevention and surveillance activities (Lub 2018; McConville and Shepherd 1992; Moores 2017).

Neighbourhood Watch was part of a broader shift in policing methods in the 1980s, following the upward trend in mechanisms of neoliberal governance, that reflected the interests of the business community and conservatism more generally (Moores 2017). In their study of Neighbourhood Watch programs in London, United Kingdom, McConville and Shepherd (1992) found that suspicion was heavily racialized, as over one-third of white respondents associated crime with Black people. Overall, the purported goal of Neighbourhood Watch programs to effect positive social change through civic surveillance is not attainable; instead, these types of programs can fracture social relations by stigmatizing segments of a community. The authors also determined that the Neighbourhood Watch model not only exploits fear of crime but can also actually increase fear of crime. The program in Heron Gate did just that (Meredith and Paquette 1992).

Heron Gate's Apartment Watch experiment was largely unsuccessful. It failed to meet the 90 percent participation threshold due in large part to lack of interest and tenant turnover. Further, surveys reveal that the perception of crime increased following the program. Survey respondents became more likely to perceive high rates of crime (an increase from 16 percent before the program to 23 percent after) and average rates of crime (an increase from 29 percent before the program to 51 percent after) in the neighbourhood, while the proportion of tenants believing that their building had a serious crime problem rose from 23 percent before the program to 35 percent after. Fewer respondents felt safe in terms of walking alone at night, with a decrease from 76 percent before the program to 56 percent after (Meredith and Paquette 1992).

The Apartment Watch program in Heron Gate illustrates how efforts at social control enhance suspicion and fear of crime. This early attempt to securitize Heron Gate contributed to stigmatizing and shaping the neighbourhood as a problematic and undesirable space. This in turn would facilitate allowing the rental community to fall into disrepair when the property was sold to TransGlobe in 2007. While crime levels have decreased over the long term, perceptions of crime have increased along with the presence of police and private surveillance programs (Meredith and Paquette 1992) and crime-centric media coverage.

From perception to narrative: Crime, violence, and the media

The stigmatization of Heron Gate has been fuelled by media coverage associating the neighbourhood with crime and violence. Black male youth of Somali heritage, in particular, have borne the brunt of crime-centric news coverage. Representations of Somali-Canadian youth in media typically relate to radicalization, terrorism, and gang violence (Jiwani and Al-Rawi 2021). These stereotypical tropes become attached to racialized neighbourhoods and diasporic communities (Kadıoğlu 2022; Loury 2021), where Blackness comes to represent "a criminal and suspect category, which belongs elsewhere, is ahistorical, is invading Canada" (McKittrick 2006, 102). Somalis in Canada feel racially stigmatized (Kusow 2001), the majority of refugees in the 1990s being single mothers who experienced public hostility and state repression (Maynard 2017). Over the last twenty years, media coverage of the Somali diaspora in Ottawa has been consistently negative and crime-focused (Saxena 2021; Xia 2020).

Heron Gate is often portrayed in media accounts as a bastion of criminal and gang activity, with significant racialized and anti-immigrant undertones. Saxena's (2021) research into media coverage of the neighbourhood found that almost half of related articles appearing in the *Ottawa Citizen* from 1986 to 2019 reported on criminal activities. A supplementary search of coverage in the *Ottawa Sun* revealed sensationalizing headlines, notably in 2013 and 2014, that described a "troubled 'hood" (Bell 2013) "plagued by gun violence" (Bell 2014b) and "rampant crime" (Bell 2014a), with "residents living in fear" (Bell 2014b). The timing of the coverage is significant, leading up to the first mass eviction in 2015–16.

Media coverage has also fixated on the lack of policing in Heron Gate. It has been referred to as a "ghetto," a term associating the neighbourhood with poverty and crime (Haynes and Hutchison 2008). The interrelated social processes of ghettoization are constituted by racial stigmatization, economic disadvantage, segregation, and social policy (Chaddha and Wilson 2008). Part of the association of Heron Gate with a ghetto relates to the design and layout of the townhome complexes, which can facilitate and hide criminal behaviour, according to the developer. Timbercreek promoted its redevelopment plans (which are explored in greater detail in Chapter 9) as providing a diverse approach to density by not segregating low-rise buildings from high-rise buildings.

Xia (2020) interviewed neighbourhood residents who identified as immigrants, and they spoke about the positive qualities and experiences within the community, including access to amenities, social networks, and community supports. She then juxtaposed these resident experiences with the negative perceptions of Heron Gate that outsiders held as demonstrated through news media and social media. Over 40 percent of 335 news articles from 2008 to 2019 focused on crime, with a spike in articles about the neighbourhood's crime problem and a desire to "clean up" the area after the evictions commenced in 2016. Posting in the online forum Reddit, a largely young, white, male user base discussed the neighbourhood in negative terms and demonstrated racist attitudes toward the Somali community, referring to the homes as "absolute shit holes" and calling the neighbourhood "little Mogadishu," "lil Somalia," and "a complete shit hole" (Xia 2020, 96–97). These racist tropes linking Somali youth with crime, and describing Heron Gate as a ghetto, contributed to a broader public narrative that the neighbourhood deserved to be demolished.

Strategic Neglect

Narratives of criminality and racial stigma enable patterns of landlord neglect. One resident I interviewed made the connection between the issues of stigma and housing, discussing how the townhomes were allowed to fall into disrepair:

> If I tell people that I live in Herongate, automatically their first response is to kind of raise eyebrows and be like, "Oh, you're from there." And I'm like, "Well yeah, it's not a bad area." I know that it has a narrative to it for sure, but that narrative, I think, is one that is based essentially in white supremacy, in a falsehood. Because there is that idea of like, "Oh, you walk there at night, you'll get snatched up or something bad will happen to you." But I've never felt unsafe walking here until 2016, until 2018, when our area started getting torn down, when things started to get really dilapidated in this area, when housing became a huge issue. It was always an issue in this area, but it became even more so of an issue. (Anonymous interview, May 27, 2021)

The cycle of stigma, neglect, dilapidation, and demolition described by this resident, who lives in a section of the neighbourhood that the Herongate Tenant Coalition believes will be the next slated for demolition, is palpable.

Property owner and landlord Timbercreek often blamed neglect on the previous landlord, TransGlobe, and was quick to highlight investments and "improvements," including repaired sidewalks, parking garages, re-faced buildings, and landscaping (McCracken 2015). Timbercreek Communities president David Melo explained that improvements "typically come with higher rents, or improved rents from when we acquired a property" (Brent 2016). Implementing aesthetic "improvements" is part of a strategy referred to as "gentrification-by-upgrading," where buildings are renovated in order to raise rents and transform communities to attract more affluent tenants with greater purchasing power (August and Walks 2018).

However, no improvements were implemented within the parcels that contained the hundreds of rental townhouse units. Instead, "rehabilitation" and "rejuvenation" efforts were directed toward the apartment buildings alongside a redevelopment strategy "to increase the number of leasable units" (Momentum Planning & Communications 2021). Even so, Timbercreek applied to the Landlord and Tenant Board for an above guideline increase (AGI) in rent to cover the costs. The largely aesthetic upgrades, which residents argued were undertaken to improve the view for the new residents in the new redevelopment, would be paid for by tenants. Meanwhile, the townhomes continued to fall into disrepair.

Ignoring work orders

Beginning in 2015, Timbercreek rolled out a steady narrative that the townhomes were "at the end of their life cycle" and "no longer viable" (Willing and van der Zwan 2018), and that it was "not economic to make patches anymore" (McCracken 2015). Tenants, however, have accused Timbercreek — and previous landlord TransGlobe — of intentionally allowing the housing to decay to the point where demolition could be justified (Egal 2018; McCracken 2015; Rockwell 2018a). As Egal (2018) notes, "Many tenants feel that Timbercreek has purposefully neglected the houses in Herongate because they knew they intended to evict the tenants, demolish the houses, and build housing that is targeted towards higher income tenants."

Suspicions are supported by internal documents obtained through FOI requests to the City of Ottawa (City of Ottawa 2018-621, 2018-629,

2018-744). By-Law and Regulatory Services is charged with enforcing property standards and property maintenance bylaws. Bylaw officers investigate complaints and issue work orders to property owners found in violation of the bylaws. The FOI disclosures demonstrate a disproportionate number of tenant complaints and city-issued work orders in Heron Gate, including in the demoviction zone. According to the Herongate Tenant Coalition (2018a), the landlord has "weaponized" purposeful neglect as an instrument of mass displacement.

Saido Gashan and Abdullahi Ali, who were members of the Herongate Tenant Coalition, lived in a townhome slated for demolition. They had moved into the unit in 2015, after their previous home in Heron Gate was scheduled for demolition in 2016. Their new home was in considerable disrepair with a number of serious defects. Flooding in the basement had rendered the area unliveable, destroyed their possessions, and created a health hazard for their large, multigenerational family. Gashan and Ali struggled relentlessly to get repairs, submitting six written communications asking Timbercreek to fix the leak in their home between May 2017 and June 2018 (Gashan and Ali v. Timbercreek 2019; see also City of Ottawa 2018-632).

Timbercreek was very slow or altogether unresponsive in carrying out maintenance and repairs requested by Heron Gate residents. Tenants often had to reach out to the municipal bylaw office. When major flooding occurred in their basement in July 2018, Gashan and Ali escalated the issue to the bylaw office. Shortly thereafter, in August, a bylaw officer informed Ali that he had issued a repair order to Timbercreek. After a few weeks of inaction, Ali followed up with the bylaw officer, who said Timbercreek had indicated the repairs had been made. Neal Rockwell (2021, 146), an independent journalist who worked closely with the community in 2018 and documented residents' experiences of the mass eviction, notes that Ali was incensed:

> For Abdullahi this highlighted again the way he and the other residents were never listened to, never taken seriously, no matter how reasonable their position, no matter how many people corroborated the story, versus the company [Timbercreek], which, by any one official, with some power or authority, was always heard, regardless of how implausible their account might be.

Neglect falls within the wider strategic repertoire of financialized landlords seeking to squeeze profits by not paying for repairs and maintenance as well as purposefully neglecting homes to the point of justifying their demolition to intensify the property. Yet there remains a more insidious part of this story that further demonstrates how mechanisms of municipal governance facilitate landlord misconduct that contributes to their unjust enrichment.

A loophole for landlords

What Gashan and Ali did not know was that after the bylaw officer issued the order for Timbercreek to repair their basement, the landlord found a way to get out of it. FOI requests I filed with the City of Ottawa while working with the Herongate Tenant Coalition led to the discovery of a quasi-judicial board, the PSLAC, which allows landlords to appeal repair work orders issued by By-Law and Regulatory Services without informing tenants (City of Ottawa 2018-744). This type of committee, which takes different forms in different municipalities, was legislated into existence in Ontario through the 1997 *Tenant Protection Act* rental housing deregulation regime. One tenant organizer who was familiar with the Toronto version of the PSLAC referred to it as "law optional," meaning that landlords could use this municipal mechanism to renege on maintenance obligations required by law.

Obtaining this FOI disclosure was significant because it revealed that instead of abiding by the repair order, Timbercreek had appealed to the PSLAC in a letter dated August 20, and a hearing was held on September 19. An audio recording of the hearing was included in the disclosure. In the hearing, a Timbercreek employee offered a creative version of alternative events, or in legal terms "fraudulently misrepresented the facts," regarding fulfilling the work order at the townhome inhabited by Gashan and Ali (Payne 2019). But the issue was not simply a matter of a bylaw officer taking the word of a landlord over a tenant. The PSLAC hearing only came to light as a result of the FOI disclosure. Gashan and Ali were not notified about the hearing or the decision.

Gashan and Ali have brought a small claims lawsuit against Timbercreek for unjust enrichment, seeking $25,000 plus legal costs. The statement of claim draws from the materials in the FOI disclosure, including the letter and the audio recording of the hearing. During the hearing, Timbercreek's John Loubser claimed there was no history of water infiltration in the basement and that the tenants had never complained of such. He further suggested that Gashan and Ali made the claim because of their involvement with the Herongate Tenant Coalition, and said the two "are very resistant to having to move." Loubser depicted Gashan and Ali's actions — attempting to have their home maintained and repaired — as malicious: "Our perspective is that they have, they have gone this route, called Property Standards and caused this order to issue under the circumstances, simply to use the Property Standards By-Law to be punitive against us when there's really no merit to it" (City of Ottawa 2018-744).

But shortly after blaming the disrepair on the tenants' desire to stay there, Loubser revealed the truth, explaining, "There are a number of homes in this area that are literally not viable from an economic perspective. It doesn't make sense to repair them at this point." Timbercreek noted it would cost over $12,000 to repair the home's foundation, an unreasonable consideration given the home had been slated for demolition. What is missing from this narrative, and what would surely have been raised if the tenants were given the opportunity to participate in the hearing, is that requests for maintenance and repairs had been submitted over a year earlier with no response. The audio recording captures the jovial tone of the hearing and the general agreement among the PSLAC members that Timbercreek had acted appropriately.

The proceedings of the PSLAC hearing, as well as the very existence of a municipal committee that can overrule repair orders without the knowledge of tenants, is demonstrative of a municipal-developer nexus and clear class hierarchy in the realm of property relations. This hierarchy is deepened when tenants are racialized and live in a stigmatized neighbourhood, as demonstrated informally (repair requests ignored) and formally (repair requests successfully appealed). Heron Gate tenants were forced to live in unliveable conditions, their requests for maintenance and repairs ignored, even though the landlord is legally obligated to make repairs as set out in the tenancy agreement. According to Gashan and Ali's statement of claim, Timbercreek interfered with their freedom of speech and association by alleging that they were part of a Herongate Tenant Coalition conspiracy. In addition, Timbercreek "fraudulently misrepresented material facts before the PSLAC" and "prioritized their commercial interests over their statutory duties" to their tenants (Gashan and Ali v. Timbercreek 2019, 19). The statement of claim further alleges that Timbercreek "displayed reckless disregard for the Plaintiffs' safety, health and overall wellbeing" and "allowed the Plaintiffs' Unit to dilapidate for the purpose of encouraging them to find alternative housing" (Gashan and Ali v. Timbercreek 2019, 18–19).

Patterns of systemic and purposeful neglect illuminate how real estate actors benefit from deregulation of the housing market as well as manipulate state bodies to allow them to renege on maintenance obligations. Timbercreek's subversion of the municipal property standards and bylaw regime reveals an overarching strategy of maintaining disrepair, at the expense of the most marginalized, in order to later justify evictions and demolitions. The Heron Gate community has a strong sense of belonging based on mutual difference and shared struggle, despite facing structural discrimination, stigma, and racialized poverty, which have served an outsider narrative that the neighbourhood deserves to be torn down. But this story could not be told without the financialization of rental housing, a rising phenomenon where financialized real estate investment firms purchase and reposition rental housing stock.

CHAPTER 6

Heron Gate and the Financialization of Rental Housing

Real estate investment firms deploy various strategies to enhance revenue streams, which in the case of Heron Gate Village and its landlord Timbercreek Asset Management (now Hazelview Investments) includes targeted gentrification, mass demovictions, the destruction of an ethnoracial enclave, and the displacement of hundreds of households. The company's development into one of Canada's largest financialized landlords, owning billions of dollars worth of property, is not an anomaly. Rather, the rise of financialized real estate firms and the enrichment of their executives and shareholders is part of a story of how racialized property relations evolve in a settler colonial society — how typically white, male Canadians profit from buying, selling, owning, and renting Indigenous land and displacing lower-income tenants from affordable housing. Sometimes gentrification entails the destruction of distinct urban spaces, including ethnoracial enclaves such as Heron Gate. Under what Farha (2018) calls the "corporate capture of housing," investors do not think of buildings and neighbourhoods in these terms; instead, they approach rental housing and urban space in terms of return on investment. This chapter documents how Heron Gate tenants came to live under financialized ownership and then eviction, demolition, and redevelopment.

The Ownership Trajectory of Heron Gate

One day while waiting for a small claims court hearing related to Heron Gate, Herongate Tenant Coalition member Josh Hawley and I decided to visit Ottawa's Land Registry Office, which is situated in the courthouse building. There, we retrieved a survey document showing how the Heron

Gate parcels (or "blocks") were divided and sold. The survey certificate and map reveal that in August 1965, Minto Construction, under president Irving Greenberg, took ownership of these parcels of land. The ownership certificates demonstrate the ease with which settler entities transfer Algonquin land between themselves, as well as the role of the survey and map in carving up Indigenous land for development (Blomley 2003). With an ownership monopoly over the neighbourhood, Minto designed and developed a planned community with thousands of rental units. Heron Gate Village was completed in the late 1960s, at a time when the larger Alta Vista area was undergoing urbanization. Minto, still owned by the Greenbergs, one of Canada's wealthiest families, has since grown into a real estate empire worth over $4 billion (Hawley 2018; Minto 2021).

Following Irving Greenberg's death in 1991, Minto's properties were split between the family heirs. Irving's son Dan Greenberg created Otnim Properties ("Otnim" is Minto backwards) and took ownership of the Heron Gate properties from 1997 until the neighbourhood was sold in 2007 to Daniel Drimmer's TransGlobe — or so it seemed. The new ownership structure was far from straightforward. The Herongate Tenant Coalition and its research team have been relentless in identifying Heron Gate's owners, and in 2018 Hawley outlined the ownership web in an article in *The Leveller*. Heron Gate had been branded as a TransGlobe property, but documents obtained from the Land Registry Office do not name TransGlobe as an owner. Instead, ownership fell to a numerically ordered series of companies under Kanco Heron Gate Ltd. that corresponded with the neighbourhood's various parcels of land (e.g., Kanco Heron Gate-1, Kanco Heron Gate-2). The ownership structure was linked to Drimmer's family real estate dynasty, which Hawley (2018) described as a "serpentine ownership network of corporations, partnerships and trusts."

From 2006 to 2009, TransGlobe gobbled up properties across the country, becoming Canada's third-largest landlord. Following the acquisition of Heron Gate, living conditions in the complex sharply declined as the landlord ignored maintenance orders. Other cost-saving and profit-enhancing strategies created conditions that led CBC's *Marketplace* to name TransGlobe as "one of Canada's worst landlords" (CBC 2012). With this grim reputation hanging over it, TransGlobe transformed in 2010 into a real estate investment trust (REIT) and quickly redistributed its properties to other financialized real estate investment firms, including some controlled by Drimmer (most notably Starlight Investments). In 2012 the TransGlobe brand ceased to exist, and Heron Gate came under new ownership. Timbercreek Asset Management purchased Heron Gate in a set of acquisitions in 2012 and 2013 for under $200 million.

But the Herongate Tenant Coalition's research has revealed that, like TransGlobe, Timbercreek was not named on the deed. Instead, the listed owners were Mustang Equities Inc., TC Core GP, and TC Core LP. Hawley also discovered that in 2012 Mustang Equities had formed partnerships with companies affiliated with Drimmer: Mustang-Master LP, Mustang DDAP Partnership, and DD Mustang Holdings GP. These partnerships all have the same business address in Toronto, which is shared by many of Drimmer's other companies. "Mustang," according to Hawley (2018), links Drimmer with Timbercreek cofounder Ugo Bizzarri; the two graduated in 1993 from the Ivey Business School at Western University, whose athletics teams are branded the Mustangs. Timbercreek's other cofounder, Blair Tamblyn, also graduated from Western University in 1994.

The insidious connections between real estate juggernauts provide valuable insight into the intricacies of the corporate capture and financialization of rental housing, but they also form part of a story of property and privilege in settler colonial Canada. Success in settler society is often attributed to hard work, a formal postsecondary education, and savvy entrepreneurship, contributing to a powerful and persistent myth about entrepreneurial achievement and property ownership. Dependent on the ongoing theft and exploitation of Indigenous lands, settler business ventures and the attempted perfecting of property relations through real estate are central to settler life, flourishment, and value. The executives that form real estate investment entities and gentrify lower-income, working-class, immigrant neighbourhoods have capitalized on the social, political, and economic conditions and policies of an evolving settler colonial society centred on, and driven by, racialized property relations.

Housing Policy, Corporate Capture, and Financialized Gentrification

The situation at Heron Gate provides insight into how the financialization of rental housing and settler colonial property relations drive the displacement and replacement of lower-income, racialized households.

Financialized real estate emerged from the significant state withdrawal from the provision of housing and related responsibilities in the 1990s (Suttor 2016). Although finance has always been an important underpinning of economic activity, the phenomenon of financialization has increasingly garnered scholarly attention since then. Sawyer (2013, 315) defines financialization as "the expanded role and volume of financial markets, institutions and agents, and products within global circuits of capital since the decommodification of world money in 1971." That was the year in which the Keynesian consensus collapsed, triggering the ascent and institutionalization of finance-led growth and birthing the modern neoliberal era. Rutland (2010) identifies four distinct approaches to financialization, including the increasing significance of financial institutions and activities, corporate drive to maximize shareholder value, a shift from interest-bearing to fictitious capital, and embeddedness of worldwide economic activities in financial markets.

Real estate and the housing sector have become primary targets for financialization and neoliberal policy prescriptions. Urban restructuring has dramatically altered the housing landscape in Canada since the 1990s (Suttor 2009), reaching a critical juncture in 1993 when federal legislation slashed social housing funding, enabled investment firm access to the real estate market, and downloaded responsibilities for housing to lower levels of government (August and Walks 2018; Walks and Clifford 2015). In Ontario additional deregulation occurred in 1997 with the *Tenant Protection Act* (precursor to the 2006 *Residential Tenancies Act*), which removed rent controls and further facilitated investor access to multifamily rental housing. These conditions allowed new players to enter the housing investment field, where real estate investment firms could capture and financialize the rental housing sector. Meanwhile, widening wealth gaps and decreasing purchasing power have rendered lower-income renter households more vulnerable to gentrification and displacement, exacerbated by longstanding housing and municipal policies that emphasize homeownership, single-detached dwellings, and suburban sprawl.

A United Nations Human Rights Council (2017) report focusing on the financialization of housing — penned by Leilani Farha, the special rapporteur on adequate housing — traces the rise and infusion of finance capital into the housing sector with the advent of mortgage-backed securities in the 1980s. Financial entities grouped multiple mortgages into portfolios, which they then sold to investors. Alongside the deregulation of housing markets, mortgage-backed securities led to the "increased use of housing as an investment asset integrated in a globalized financial market" (Rolnik 2013, 1,059). Walks and Clifford (2015, 1,639) describe "an unsung policy of encouraging private sector investors to become landlords to fill the demand for rental housing," resulting in the growing domination of the Canadian rental housing market by landlord-investors.

The corporate capture of housing signals an intensification of real estate investment activity by large corporate and financialized firms. Corporate finance is rapidly commodifying the housing sector, which provides a site for significant investments of global capital, a source of security for financial instruments traded on global markets, and a means of accumulating wealth (Aalbers 2016; Human Rights Council 2017). According to the United Nations Human Rights Council (2017, 3) report, the housing sector is at the centre of a "historical structural transformation in global investment and the economies of the industrialized world." Financialized real estate investment firms have capitalized on the state's withdrawal from housing provision to capture and redevelop rental housing stock. In discussing the financialization of rental housing in Germany, Wijburg, Aalbers, and Heeg (2018) identify a shift from speculation and short-term investments to a focus on stable cash flows in long-term investments. The financialization of rental housing was initially characterized by a "buy low, sell high" ethos involving the acquisition of land and real estate by private equity and other investment funds; however, the financialization of rental housing increasingly involves a renewed interest in rentiership that treats real estate as a long-term investment strategy (Aalbers 2019, 381).

Rental housing is an important node for financialization projects on a global scale (Fields 2017) and recent research shines a spotlight on the impacts. Lewis (2022) documents the uneven racialized impacts and violence of housing financialization in Toronto (see also Fields and Raymond 2021), important research that sheds light on the inequality inherent in housing systems in settler colonial countries. Scholarship on Black geographies (Bledsoe and Wright 2018; Hawthorne 2019) is crucial to consider alongside housing research as housing systems are implicated in racial segregation, spatial inequality, place identity and belonging, and resistance. Fields and Uffer's (2016) study of private equity real estate investment in New York and Berlin and August and Walks' (2018) examination of financialized landlords in Toronto highlight not only the vulnerabilities of lower-income renters but also the gentrification and reconfiguration of buildings and neighbourhoods. Financialized gentrification results when a financialized real estate investment firm takes over an apartment building or neighbourhood and pushes out lower-income tenants (typically racialized and marginalized populations) to make way for higher-income tenants (and typically white) (Crosby 2020a). In this way, financialized gentrification produces and reproduces social and spatial inequalities, entailing "unscrupulous demographic engineering in search of profits: replacing poor and vulnerable people with those who possess greater purchasing power" (Farha 2018).

"The Apartment as Saviour": Strategies of Real Estate Investment Firms

Real estate investing is an increasingly favoured method of building wealth, and a variety of financial vehicles are available for individual and institutional investors in both the public and private sphere. REITs and private equity funds have emerged as powerful players in the market, along with other asset management companies and investment firms. Whereas single-family properties have long been a primary focus for individual investors, institutional investors have targeted multifamily rental properties. In their own words:

> Multi-unit residential has moved to the top of the preferred property portfolio for a growing number of real estate organizations across Canada. What was once an asset class dominated by smaller privately-owned firms has now also become the domain of major institutional investors, REITs and private equity funds. Predictable yields, record returns and strong cash flows have made apartments one of the most reliable property classes. (Canadian Real Estate Forums 2019)

Financialized real estate actors are cornering the rental housing market in Canada. A 2015 survey by *Canadian Apartment* magazine demonstrates that the largest players — those owning over 7,500 suites — are predominantly REITs and other investment and asset management firms (remi Network 2015). August's (2020) research provides more recent numbers, documenting that all but three of the top sixteen apartment landlords owning over 7,500 suites in Canada are financialized real estate investment firms, according to 2017 figures. Recent and ongoing conversions of private equity funds and private corporations to REITs, as well as REIT consolidations, also show how institutionalized investment players are taking over the rental housing market. One presentation at the 2017 Canadian Apartment Investment Conference touted "the apartment as saviour" and outlined that the "multi-residential transaction volume" in Canada has remained steady at $3.5 billion to $4.5 billion annually since 2011 (Chandler 2017). The same presentation showed that of the top ten buyers of apartments in Canada over the previous twenty-four months, all but one were real estate investment and asset management firms. The top five included CAPREIT, Starlight Investments, Northview Apartment REIT, Skyline Apartment REIT, and Timbercreek Asset Management, collectively accounting for over $3 billion in purchases (Chandler 2017).

Private equity funds, REITs, and asset management firms have encouraged a renewed interest in rentiership (Blakeley 2019; Christophers 2020). Up to half of Canadians are investing in stocks (Randall 2021), and financial entities have enticed the everyday individual retail investor to benefit from sustainable and growing rent profits without the hassle of engaging in property management themselves. Investing in companies that own and manage apartment units is an easy way for the average settler citizen to participate in the real estate market and profit from the development of Indigenous land. The investment objectives of real estate investment firms are relatively uniform, as captured in the 2018 annual report of Canada's largest landlord, CAPREIT: to provide shareholders with long-term, stable, and predictable monthly cash distributions. In order to provide regular cash flow to investors, investment firms engage in aggressive property management strategies to raise rents, reduce expenses, and exploit efficiencies to maximize asset value. These strategies can lead to processes of gentrification and displacement.

Financialized gentrification can occur when a financial instrument or entity subjects a property to real estate investment strategies to "add" or "create value" to the obtained assets. According to August and Walks (2018), financialized landlords engage in two key strategies to extract value from their properties. First, they "squeeze" profits from tenants by implementing cost-cutting measures, efficiency upgrades, new costs for tenants (such as ancillary fees and submetering utilities), and rent increases, which in Ontario can include raising rent above the annual maximum amount (also known as AGI for "above guideline increase") if a formal request to the Landlord and Tenant Board is approved. The AGI is part of the larger trend of neoliberal deregulation of rental housing initiated in Ontario with the 1997 *Tenant Protection Act* along with vacancy decontrol, which permits landlords to raise the rent of a vacant unit by an unlimited amount (Crosby 2020b). Second, financialized landlords engage in a strategy of "repositioning" buildings, what August and Walks (2018) refer to as "gentrification-by-upgrading," to transform their tenant base. Repositioning involves converting affordable units into luxury suites in coveted market locations, with sharp increases in rents. In an annual report, Ottawa-based InterRent REIT (2014, 5) put it plainly, stating the trust seeks to "grow the rental revenue base organically while at the same time improving its stability by removing undesirable tenants and implementing policies and processes to attract more desirable tenants." Enabled by state policies and legislation, REITs and financialized property management firms have thrived as apartment investors, bringing with them evolving sets of accumulation strategies that have accelerated the gentrification of lower-income buildings and neighbourhoods, as well as eroded affordable housing stock, in Canadian cities.

Timbercreek: "Actively Creating Value"

Toronto-based Timbercreek has emerged as a major player in Canada's rental market. Founded in 1999, the company has steadily evolved from its made-in-Ontario roots to a global player with asset management offices on three continents. Timbercreek manages billions of dollars of property and other assets through public securities and lending, development, and property management operations (Wong 2019).

Just as the ownership structure of real estate firms like Timbercreek is opaque, so too are the investment entities and vehicles with which it conducts business. The firm offers a wide variety of public and private investment vehicles marketed around the firm's slogan, "actively creating value." An investment brochure describes the company as an "active investor, owner and manager of global real estate and related assets focused on delivering sustainable and growing returns" to investors (Timbercreek 2019a). An older version of its website elaborates on the firm's mandate and vision:

We maximize value by employing a value-oriented investment philosophy combined with an active, hands-on asset management platform, to identify opportunities that will generate predictable and sustainable long-term cash flow. We have earned a reputation for providing conservatively managed, risk-averse investment opportunities for both retail and institutional investors. (Timbercreek Asset Management 2019b)

Timbercreek boasts as its "core competency" the identification of high-quality real estate opportunities by applying "bricks-and-mortar knowledge," which refers to the "ability to accurately value cash flows based on a comprehensive analysis of the quality and sustainability of a property's current and future revenue streams" (Timbercreek Asset Management 2019b, 2019c). Investments target private equity (investing directly in real estate), private debt (investing in mortgages and other debt secured by real estate), and public securities (investing globally in publicly traded companies that own investment-grade real estate) (Timbercreek Asset Management 2019d). In its own words, Timbercreek "focuses on accessing stable, inflation-hedged cash flow by investing in real estate directly, investing in debt secured by real estate and investing in companies (publicly and privately) that own real estate" (Timbercreek Asset Management 2019a). The firm claims to have one of the leading public real estate securities teams and access to a network of co-investors that includes Canada's largest funds and leading real estate private equity teams (IPE Real Assets 2015). Its multipronged approach to investing in and managing real estate assets and debt further set it apart from traditional landlords and developers.

Financialized real estate investment firms are always looking for new market sectors to corner (August 2020). For example, Timbercreek's 2019 market outlook report identifies industrial REITs and Canada's senior housing sector as promising opportunities, along with diversification beyond a traditional residential tenant base to include data centres, cell towers, hotels, and casinos (Timbercreek Investment Management Inc. 2019; Wong 2019). Corrado Russo, Timbercreek's senior managing director of investments and global head of securities, expressed the logic behind targeting property with a single tenant: "Your rent is very close to 100 per cent profit, and while you typically get less rent, there's no risk to expenses going up longer term so it behaves like a bond but a bond that gives you equity-type returns because you still have the ability to release that space at higher rents as market rents go up" (Wong 2019). Timbercreek engages in lower-risk acquisitions that accumulate stable, high yields over the long term, maintaining the option to sell properties at a profit if market conditions are favourable. The promise of predictable profits relies on gentrification, tenant turnover, and steadily rising rents.

Underpinning the firm's "value-oriented investment philosophy" is the drive to seek out and obtain "real estate assets that [are] undervalued or had been overlooked by the market" (Threndyle 2009, 24). Bizzarri once likened the company's business strategy to a carwash involving the purchase and transformation (cleaning) of neglected (dirty) properties for high-profit resale (August and Walks 2018, 132). Tamblyn also used the carwash approach, when talking about Timbercreek's foray into the U.S. market with a $100-million multiresidential value-added fund: "On the front end, you put in a multi-res asset that, in our view, has not been operated to its fullest potential, and about two and a half years later, it comes out the other end looking squeaky clean and ready for an institutional buyer to acquire it" (IPE Real Assets 2015). Timbercreek's strategies continue to evolve, and while flipping properties remains an option, the firm has shifted its approach in Canada as a property manager and active landlord, targeting lower-income buildings and neighbourhoods in urban areas ripe for gentrification.

Gentrification and intensification

In an interview with *Western Investor* (O'brien 2018), Russo emphasized the importance of capitalizing on displacement in the rental housing sector: "Large cities in Canada are currently experiencing a wave of gentrification," he said, which "is creating a number of compelling opportunities for REITs to experience outsized growth and offer increasing value for investors."

His admission, which is not unique in the industry, revealed Timbercreek's predatory approach of cherry-picking and repositioning properties in gentrifying communities, which both contributes to and exacerbates gentrification-induced displacement. Companies like Timbercreek also work feverishly to produce gentrification conditions through the targeting of certain areas, typically home to lower-income and racialized populations; the potential for "outsized growth" and "increasing value for investors" outweighs any concerns about the expulsion of tenants. As Timbercreek continues to capture what it deems as "undervalued" properties throughout North America — totalling over two hundred multi-unit buildings with some 23,000 suites (Timbercreek Asset Management 2019e) — it is not merely a passive player taking advantage of shifting demographic trends. On the contrary, Timbercreek is an active gentrifying agent engaging in demographic restructuring and sociospatial reengineering of buildings and neighbourhoods as it fulfills promises to investors.

Timbercreek officials explicitly set themselves apart from traditional private firms, noting that small owners do "a less than exemplary job of squeezing all the value out of their buildings" (Threndyle 2009). The firm avoids investing in "developers and homebuilders which don't offer that opportunity for recurring revenue," Russo has said (Valiquette 2019). Instead, its strategy is predicated on "hunt(ing) for companies" in "trophy markets" and "finding gems" that "are under-loved and underpriced" in order to "build" value. According to Russo, "the more upside potential the better, whether that is achieved through retrofits and upgrades to the property, improvements to the lease structure or having the property rezoned for a more lucrative use" (Valiquette 2019). Improving the lease structure, which Russo enthusiastically referred to as "triple net leasing," renders tenants responsible for paying for the maintenance and operating costs of the property. For Russo, triple net leasing is a means of "really getting equity-type returns for a bond-like risk." He also noted the tactic of rezoning properties. Eviction, demolition, and rezoning are necessary precursors of a new intensification strategy for financialized real estate firms to extract "more lucrative use" from their assets.

Intensification has the potential to generate new profitable assets either by way of new-build developments on vacant land (otherwise known as infill developments) or by demolishing existing structures and redeveloping the already-built environment. Real estate investment firms and asset management companies increasingly consider what they call "site intensification" or "intensification opportunity" when acquiring properties. Conferences on real estate investment provide a unique opportunity to witness those at the upper rungs of the apartment investment industry share strategies and discuss annual returns on capital and income growth, capitalization rates, development projects, and more. At a 2017 Canadian Apartment Investment Conference session, apartment owners and investors discussed the importance of embracing intensification. "If you want to control your destiny as an apartment owner in Canada now, you have to be in the new construction game because there's not a flow of deals anymore," warned a representative from CAPREIT (McLean 2017). Following municipal planning trends, apartment investors have embraced demands for density and the intensification of urban areas. CAPREIT's president and CEO Mark Kenney simplified the approach from a property investment perspective, explaining that apartment buildings are "not being valued on cash flow. Cap rates are irrelevant in the conversation. It's all about land and speculation on density" (McLean 2017). Apartment investors' insatiable thirst for access to cheap land and high rents, as well as enhanced profits from intensification initiatives, is synergetic with the municipal governance turn toward densification and partnering with property developers.

The City of Ottawa's (2021a) New Official Plan identifies intensification as one of six "milestone new directions" or "cross-cutting issues" that will help transform the municipality into North America's most liveable mid-sized city. The plan's emphasis on intensification is illustrative here as a discourse of improvement that aims to enhance liveability through gentrification. Intensification "refers to the notion of renewal and injecting new life into existing areas of the city by allowing new generations of residents to inhabit existing neighbourhoods" (City of Ottawa 2019b, 1). In the city's view, intensification creates "opportunities for housing choice" and "will enliven existing neighbourhoods by bringing new residents to support local services" (City of Ottawa 2019b, 1). But intensification, or regeneration as it was referred to in early draft versions of the plan, is not just about creating density and augmenting neighbourhood populations.

Instead, it is about regenerating state-led gentrification and slum clearance efforts in the name of urban renewal. Urban renewal efforts are often associated with large-scale state-led demolition and displacement projects targeting the racialized urban poor (Hyra 2012; Triece 2016). Despite the mounds of research documenting urban renewal's racist underpinnings (see for example Fullilove 2016), the architects of the City of Ottawa's New Official Plan invoke improvement discourses of "renewal and injecting new life" to establish gentrification as a municipal policy direction for enhancing liveability. In the municipal imaginary, the path toward liveability involves encouraging new residents to inhabit and enliven existing neighbourhoods, to inject new life into areas where existing life is deemed to be waning.

Apartment investors latch on to municipal policy prescriptions and harness popular discourses of improvement to become major players in the urban built environment. Urban improvement discourses of intensification, revitalization, and liveability, along with not-so-subtle references to gentrification, are circulated across the major apartment investors in Canada. For instance, Timbercreek executives presented slides at real estate investment conferences in 2016, promoting the Vista Local complex as an example of the successful "intensification of an existing asset" (Tsourounis 2016).

Vista Local, situated on the site of the 2016 demoviction, is the first phase of redevelopment in Heron Gate. Bizarri (2016) said the new development would create a "vibrant and sustainable rental community, catering to a balanced target market of young professionals and active adults" by offering, among other things, "resort-style amenities." He also said it represented four hundred of an anticipated five thousand to ten thousand new-build units as part of Timbercreek's intensification efforts (McLean 2017). Not mentioned on the presentation slides is that dozens of families were evicted and their homes demolished in order to realize this intensification project. In this example of financialized gentrification, lower-income, racialized, immigrant families are not covered by Timbercreek's "balanced target market." The human element of urban redevelopment is missing not only from the images presented at real estate investment conferences but also from apartment investment executives' worldview.

While there are many circumstances in which a home can be unmade, the worldviews and actions of apartment investors — operating within the parameters of settler colonial property relations — accelerate gentrification, displacement, and ultimately domicide. Timbercreek/Hazelview claims to care about the wellbeing of the community. Its public-facing position is that its efforts to improve the community unfortunately result in the relocation (eviction and expulsion) of some tenants, a by-product of the revitalization (gentrification) of urban space: this is how neighbourhoods become liveable. The message to investors about desirable and undesirable tenants, on the other hand, boils down to the ability to pay rents. From the outset, financialized landlords consider tenants in Heron Gate and similar buildings and neighbourhoods undesirable because they pay below-market rents. Desirable tenants can pay above-market rents. At Heron Gate, Timbercreek/Hazelview desires what it refers to as "premium" tenants paying premium rents (City of Ottawa 2020-424; McCracken 2016a).

The financialized landlord purports to be colour-blind and gender-neutral. Timbercreek/Hazelview executives dispute the charge that they are destroying an ethnoracial enclave and are adamant that they do not conduct business through lenses of race and gender — much the same as many average white settler Canadians defensively deny the existence of systemic racism, despite the prevalence of anti-Black, anti-Indigenous, anti-Asian, and Islamophobic sentiment in a settler society and state built upon the pillars of colonization and white supremacy. What the Timbercreek/Hazelview executives fail to acknowledge is that, like themselves, the owners and leaders of large corporate and financialized real estate investment firms are white men of privilege, and the biggest impacts of their gentrification efforts come down on racialized women. Single mothers and primary caregivers of large, extended families rely on communities like Heron Gate for their support networks and income. Predominantly white real estate executives, city officials, and urban planners initiate and facilitate the removal of racialized households, and largely white demographics replace the existing population and benefit from gentrification and domicide. Settler colonial property relations are structured around the value of whiteness and the devaluation of nonwhite lives.

Rebranding as Hazelview

The hubris with which Timbercreek approached the mass eviction of hundreds of Heron Gate residents is part of "dehumanized housing," which is predicated on eviction, displacement, and replacement (Human Rights Council 2017, 9). Focused on investment jargon — such as returns, yields, and equity — Timbercreek's executives neither accounted for nor anticipated the backlash from tenants. The Herongate Tenant Coalition (detailed in Chapter 8) directly confronted Timbercreek executives and employees, as well as the Timbercreek brand, for their dehumanized approach to housing. Following the coalition's relentless social media, news media, and legal campaigns, Timbercreek rebranded, trying to disassociate its name from the neighbourhood.

The priorities and branding of real estate investment firms can shift over time in response to economic and social forces. Researching firms like Timbercreek in recent years has revealed some of the numerous entities with which the larger brand navigates the world of real estate investment — including Timbercreek Asset Management, Timbercreek REIT, Timbercreek Financial, Timbercreek Capital, Timbercreek Equities, Timbercreek Communities, Timbercreek Investment Management — as well as various publicly traded and private equity investment funds. The two Timbercreek cofounders split up the original Timbercreek into Timbercreek Capital (run by Tamblyn) and Hazelview Investments (run by Bizzarri). The Timbercreek entity in Heron Gate was replaced by Hazelview, but regardless of the corporate rebranding, the ownership structure of Heron Gate has not changed. The organizational entities that formed Timbercreek Asset Management and related investment vehicles, and that purchased and hold fee simple title to the Heron Gate lands, maintain their original legal structure and names. This is documented in the MOU for the Heron Gate Official Plan Amendment between the City of Ottawa and "Mustang Equities Inc. & TC Core LP (together, 'Hazelview')" (City of Ottawa 2021c).

In a press release, Timbercreek detailed how Hazelview Investments was a rebranding of an entity known as Timbercreek Equities: "Timbercreek will continue to offer commercial real estate financing solutions, while Hazelview will continue to focus on providing opportunities for investors to access real estate both privately and publicly, and will continue to operate its property management service, now known as Hazelview Properties" (Timbercreek Asset Management 2020). Thus, Timbercreek exited the multiresidential property development and management game to focus on commercial real estate lending and investment through two principal entities: Timbercreek Capital and Timbercreek Financial. Taking on residential development and management, Hazelview has adapted the original Timbercreek mission statement, maintaining a commitment to "creating value" and investing in "sustainable long-term cash flow" (Timbercreek Asset Management 2020). As Timbercreek once sought "hidden gems" like Heron Gate, Hazelview is on the lookout for "hidden opportunities," according to its website, which is about "getting the most value by digging a little deeper to uncover opportunities for growth that others missed" (Hazelview Investments 2022).

CHAPTER 7

Demoviction 2016: Domicide and Redevelopment in Heron Gate

For the initial phase of demoviction at Heron Gate, ruling actors recast eviction as "relocation," and redevelopment as "renewal." The marketing for the Vista Local first phase of redevelopment, in particular, concentrated on "liveable homes" (City of Ottawa 2018) and "resort-style living" (Shaw 2017), as a way to align Heron Gate with the predominantly white, affluent Alta Vista neighbourhood to the north. Drawing from materials produced in the context of the 2016 demoviction and Vista Local redevelopment, this chapter shows how efforts to harmonize the neighbourhood with Alta Vista necessitate erasing Black and Brown residents and replacing them with white ones.

Revitalization and Relocation

When neglect and disrepair take hold in a neighbourhood, racial and territorial stigmatization can serve to justify intervention in the form of revitalization initiatives (August 2014a; Horgan 2018). For Burns and Berbary (2021, 2), "revitalization and renewal suggest a progressive upscaling and improvement that conceals legacies of displacement. Revitalization is, in many instances, a friendlier way of describing gentrification in progress." Research on revitalization initiatives in Toronto's Regent Park provides insight into what has happened in Heron Gate. James (2010) argues that moral regulation efforts targeting the urban poor and working class, in particular racialized minorities and newcomers, have driven the revitalization planning, demolitions, and redevelopments in Regent Park. August (2014a, 1,330) argues that pro-revitalization discourse surrounding Regent Park "reinforces geographic patterns of socio-spatial polarization, in which racialized poverty is shifted away from the gentrifying core and reconcentrated in the city's increasingly stigmatized inner suburbs." In an assessment of Heron Gate, Xia (2020) argues that Timbercreek used negative stigma associated with the neighbourhood as justification for demolition and redevelopment.

As mentioned previously, in 2012 and 2013 Timbercreek purchased the Heron Gate property, consisting of multiple land parcels and hundreds of townhome and apartment units. In late 2015 eviction notices went out to fifty-four families on a block where Timbercreek deemed it was "not economic to make patches anymore" (McCracken 2015). The block, commonly referred to by Timbercreek as "HG7" (a shortened form of the land parcel's legal name, Kanco Heron Gate-7), contained eighty townhomes that were targeted for demolition.

Timbercreek hired a public relations firm to manage its brand during the process and produce communications materials related to the eviction and redevelopment. In a promotional piece on its website, Momentum Planning & Communications (2021) provides insight into Timbercreek's purchase of the property as a prime location for redevelopment:

> Timbercreek Communities acquired an ideally located piece of property in 2012 known as Heron Gate, situated in Ottawa's south end. In response to years of neglect and poor maintenance, Timbercreek Communities immediately began a comprehensive rehabilitation project and as of late 2015, had spent more than $30 M in property reinvestments. The next phase in the rejuvenation of Heron Gate was to redevelop the site to increase the number of leasable units.... The Timbercreek project fits in well with the City of Ottawa's revitalization plans of that particular section of Ottawa South.

The line about property upgrades is often used by the landlord; what is never mentioned is that they were directed at the apartment complexes, not the townhomes, which continued to be systematically neglected. The excerpt makes an explicit link between demoviction and redevelopment as revitalization, situating it within wider municipal planning aspirations to have that section of the city revitalized and gentrified.

The mass eviction of dozens of lower-income families and the demolition of their homes is an undeniably violent act. Timbercreek attempted to soften the violence of displacement and unhoming through manipulation of language, even claiming that residents were not evicted, but *relocated*, by insisting that tenants voluntarily left rather than were thrown out. However, Momentum Planning & Communications (2021) equated relocation with eviction:

> To clear the way for the demolition, Momentum Planning and Communications was mandated to initiate a Relocation and Communications Program in October 2015 to assist with the eviction of the existing tenants of HG7. The goal of this initiative was to ensure a smooth relocation process for the tenants and to manage the potential media and public impacts of the process on Timbercreek as a corporate landlord.

This short passage includes language that Timbercreek would never use publicly about the demoviction of Heron Gate: "demolition," "eviction," and even "landlord," which corporate and financialized landlords tend to eschew in favour of calling themselves "property management companies." Momentum's promotional piece provides insight into the role that public relations firms play in the gentrification and redevelopment initiatives of corporate and financialized landlords, clearing the way for mass displacement.

Intensification and Demoviction

As land becomes an increasingly scarce resource in urban settings, developers turn to intensification of built assets, which inevitably requires demolishing buildings and thus necessitates evictions. Demoviction can be conceived of as a gentrifying tactic that involves the demolition of existing housing stock for the redevelopment of new-build, higher-rent units. The mass eviction precipitating demolition facilitates an influx of new investment in construction, the destruction of rental units that typically offer below-market rents, and the reconfiguration of housing, neighbourhoods, and community demographics. In the case of financialized real estate, redevelopment serves to create a new asset class that can be easily rated and securitized to develop a new revenue stream. Intensification creates many more additional rental units than previously existed, at what Timbercreek refers to as "improved" or "premium" rents. Similar to strategies of squeezing and repositioning, demoviction is facilitated by the regulatory mechanisms of the state.

In Ontario, demoviction has no legal repercussions. The 2006 *Residential Tenancies Act* incorporates safeguards for tenants threatened by renoviction; Section 53(3) in the legislation stipulates that a tenant has the right to return to a renovated unit at the original rental price, although this seems to rarely happen in practice (Diwan et al. 2021). On the other hand, under Section 52, where a landlord terminates a tenancy for purposes of demolition, no such allowance is granted. This legal mechanism allows landlords to systematically neglect their properties to the point where they can claim the buildings are "no longer viable," where "viability" really refers to profitability.

In addition to provincial regulatory mechanisms that disadvantage tenants and benefit landlords, municipal actors and mechanisms of governance play a significant role in facilitating and legitimating development-induced displacement. For example, city officials openly supported the redevelopment of Heron Gate. Jean Cloutier, city councillor for the ward that includes Heron Gate, expressed publicly that he was "a little bit relieved that it's finally happening," adding that Heron Gate needs some "tender loving care" (McCracken 2015). In tandem with Timbercreek's vernacular of "improved rents," Cloutier spoke of "improvement" and how the "development will benefit the entire community" (McCracken 2015). He has also come under fire for his financial ties to developers (Osman and Chianello 2018; Willing 2018).

Public officials and urban planning professionals are also implicated in facilitating gentrification-induced displacement through, for instance, permitting the rezoning of properties such as Heron Gate for "more lucrative use" (Shaw 2017). In its first phase of intensification in Heron Gate, Timbercreek was allowed to forego restrictions on height and parking spaces (Shaw 2017), which is a common exception granted to property developers but one that demonstrates the degree of influence that developers have in housing governance (Robin 2018). Large landlords like Timbercreek also receive significant sums from city coffers. Hussein and Hawley (2021, 149) documented the public money that Timbercreek receives as a landlord, noting that it received over $1.7 million in "municipally funded grants" from the City of Ottawa from 2012 to 2019, the details of which are not explicitly defined. Despite receiving hefty amounts of public funding, Timbercreek "has failed to do regular maintenance and repairs, particularly to their Herongate properties, even when issued work orders by the City" (Egal and Hawley 2018). By granting demolition permits, allowing properties to be rezoned, and approving larger redevelopment plans — such as the Official Plan Amendment in 2021 that greenlit the demolition of 559 additional Heron Gate homes — the City of Ottawa is an active agent of gentrification.

With considerable municipal support for its neighbourhood restructuring efforts, Timbercreek set out to demolish dozens of townhouses in the first round of demovictions in 2016. Over fifty families were displaced, without objection or opposition from the city. But before the townhomes were destroyed, the Ottawa Police Service moved in to conduct urban warfare training.

The "Bombing of Heron Gate"

Capitalizing on the opportunity created by the evictions, the Ottawa Police Service's tactical unit used the abandoned buildings to train with firearms and explosives in an urban setting. In addition to blowing holes in the former homes — what the Herongate Tenant Coalition (2021a) referred to as the "bombing of Herongate" — the police staged a public relations stunt by inviting local media to film tactical officers destroying the homes. "It's not every day that we have a venue like this where we can actually, for lack of better words, destroy it," explained Staff Sergeant Paul Burnett (CTV Ottawa 2016). The Ontario Provincial Police's tactics and rescue unit was also invited to participate. Police set up numerous cameras inside and outside of homes to capture the explosions, as walls, doors, and windows were blown to bits. Documents obtained through an FOI disclosure outline a number of types of trainings and occurrence dates (Ottawa Police Service 19-494). One of the scenarios included an "emotionally disturbed person," raising the question of why police would consider deploying the tactical unit for somebody experiencing a mental health crisis.

In a video segment produced by CTV Ottawa (2016), officers boasted about the accuracy of their detonations, showing how they destroyed a room's entry points yet spared a light fixture. Addressing the dehumanizing situation, a post on the Herongate Tenant Coalition's (2021a) website draws attention to the video evidence of the landlord neglect that enabled these buildings to be emptied and bombed: "The police were happy to show off for the cameras. The media got to show homegrown explosions on the 6 o'clock news. The child's art on the fridge door tells a different story. The water damage and mould around the ceiling fan, a reminder of strategic neglect."

Local officials cheered on the bombing. Councillor Cloutier called the presence of the tactical unit in the neighbourhood a "win-win" for police, Timbercreek, and tenants, as "police presence gives a layer of security to the area" (McCracken 2016b). Cloutier also noted that it was Timbercreek who reached out to the Ottawa police, offering use of the space and handing over the keys to several units. Apparently lost on the police, the landlord, and the city councillor is how extremely distasteful it is to bomb the former homes of refugees who have escaped war zones.

The police operation rendered material the heavy symbolic and emotional harms of domicide, where homes deemed unviable were pre-destroyed by police after the residents had been evicted and dispersed. Demoviction at Heron Gate entails both the social destruction of home (the unmaking of home and community) and the material destruction of home (demolition). The bombing of Heron Gate sandwiches a third layer of violence in between, where the physical home is desecrated in the name of security by coercive agents of the state.

Vista Local: Revitalization and Resort-Style Living

Following the 2016 eviction, bombing, and demolition of eighty townhomes in Heron Gate, Timbercreek submitted a proposal to build what local residents dubbed "the three monsters" (McCracken 2016a) — three large, six-storey complexes containing hundreds of rental units covering 1.64 hectares. The new Vista Local complex, which Timbercreek boasted would offer resort-style living, sharply contrasted with the eighty townhomes that previously stood on the site (Shaw 2017). This redevelopment was the first phase of neighbourhood revitalization and the construction of "liveable homes" in the neighbourhood (City of Ottawa 2018).

In its proposal, Timbercreek pitched the redevelopment as embracing the trendiest aspects of urban improvement (including environmental sustainability and social responsibility), prioritizing pedestrians and supporting multimodal forms of transportation, strengthening the quality of open spaces, and so on. "Vista Local is part of a larger revitalization plan that will support a wide range of household structures, affordability, striking greenspace and resident interests," noted a press release (Timbercreek Asset Management 2019f). These broader appeals to the liveable city both mask and promote gentrification. Developers have welcomed discourses of urban improvement mobilized by Canadian cities, and developer plans are in lockstep with municipal ones, including the City of Ottawa's New Official Plan. Some of the earliest planning efforts for the Heron Gate redevelopment are worth recounting further here, along with ongoing efforts to rebrand and market the neighbourhood to non–Heron Gate demographics; critically, the landlord has undertaken considerable efforts to "harmonize" its Heron Gate holdings with the wealthier, predominantly white Alta Vista neighbourhood (Timbercreek Asset Management 2016a; City of Ottawa 2021d, 5).

Harmonization as gentrification

While pitching environmental sustainability, social responsibility, and other liveability niceties, Timbercreek's marketing materials reveal the overarching desire to align Heron Gate with Alta Vista. The first two planning principles in Timbercreek's July 2016 design brief for HG7 are to "(re)shape a vibrant community identity that enhances the quality of life for all" and to "recognize the future for Heron Gate as a unique, diverse and healthy sense of place" (Timbercreek Asset Management 2016a, v).

Within this discursive rhetoric of urban improvement emphasizing quality of life in a healthy community, there is an implied message that the existing neighbourhood, prior to redevelopment, is devoid of vitality. The existing community identity — shaped around an immigrant, ethnic neighbourhood — needs to be "(re)shaped" in order to make the area viable again. This is only made possible through demoviction, revitalization, and the replacement of a largely Black and Brown community with a white one. The renditions of a redeveloped Heron Gate evidence this, showcasing images of white residents enjoying the neighbourhood. In landlord imagery and in the broader settler imaginary of improving property, whiteness and whitespace are what will make the neighbourhood vibrant and healthy again.

The image of an older white couple dominates the illustration for Timbercreek's first planning principle, "to (re)shape a vibrant community identity," and is likely to appeal to the affluent, predominantly white, aging, retiree in Alta Vista. This Alta Vista homeowner may be looking to relinquish responsibility for maintaining their property and sell their home in order to cover retirement living expenses. The Vista

Local redevelopment offers Alta Vista retirees with accumulated wealth resort-style services and amenities that cater to their particular quality of life, without them having to move far. The apartment ads on the Vista Local website reflect this, tempting would-be renters to "choose freedom," and embrace "recreation over renovation" in their new "liveable homes" for a rent more than double what previous tenants on the site used to pay (Vista Local 2019; Hazelview Properties 2021).

Timbercreek's strategy to remake Heron Gate part of Alta Vista is further made clear in its design brief. Within, the stated vision for the neighbourhood is as follows: "Heron Gate is a complete, vibrant and sustainable community that enhances the quality of life for new, as well as long-time residents, while harmonizing with surrounding communities" (Timbercreek Asset Management 2016a). As one Herongate Tenant Coalition organizer put it: "It is as though ... the small, primarily Black population here, the Muslim population, it's not in harmony, I guess, with Alta Vista" (Anonymous interview, June 7, 2021).

The plan of harmonizing Heron Gate with surrounding communities was made clear by the landlord in media interviews. For example, Timbercreek's Greg Rogers, who helped lead the 2016 and 2018 mass evictions, expressly stated that "the intent is to bring the community in line more with what's north" (McCracken 2017). With the demolition of the HG7 townhomes and their replacement with resort-style suites, Timbercreek officials trumpeted Vista Local as embodying "a completely different product than what used to exist on this site" (Shaw 2017). At a September 2016 "public meeting consultation" attended mainly by Alta Vista homeowners, Rogers pitched the resort-style accommodations as a way to "build an alignment" between Heron Gate and Alta Vista, where the price of rent will "reflect the premium nature of the community" (McCracken 2016a).

The only thing standing in the way of aligning Herongate with Alta Vista was the so-called "'mess' to the west, that is, the other Timbercreek-managed townhome complexes in Herongate" (McCracken 2016a). The "mess" was how an Alta Vista resident referred to Heron Gate in the public meeting (McCracken 2016a). To address this "mess," Rogers pitched Vista Local as "seed[ing] change" in the neighbourhood: "We're not investing $100 million as a bet on rent.... This is an important investment for Timbercreek and its pension plans" (McCracken 2016a). Rogers was attempting to assure Alta Vista residents that Timbercreek was serious about a long-term strategy to gentrify the neighbourhood and deal with the "mess," permanently. The firm's institutional investors depend on the long-term, sustainable yields that the "premium" rents at Vista Local will produce.

Renditions of whiteness and "liveable homes"

The idea of seeding and infusing change in the neighbourhood through the Vista Local investment is captured in a presentation slide from a January 2018 open house event organized by the City of Ottawa and Timbercreek (see Figure 7-1). In the image, the parcels of land are already emptied of homes and lives, and arrows ending in question marks stretch out from HG7 into the rest of the community. This depicts Timbercreek's intentions to spread Vista Local's resort-style living throughout the neighbourhood — but what of the hundreds of homes and thousands of residents currently residing in the various neighbourhood parcels? They have been replaced by three question marks and what looks like a treetop, perhaps signifying a park.

The text accompanying the image suggests that "the community is improving" with the redevelopment of HG7. The replacement of the existing townhome parcels as part of this "Heron Gate renewal" is made possible by a "team effort" between Timbercreek, DIALOG (the design company), the City of Ottawa, and, as an add-on, "you – our community" (City of Ottawa 2018). Large-scale demoviction in affluent cities such as Ottawa is best served through a united front of ruling relations agents, a municipal-developer nexus that includes city planners and politicians, urban designers, and landlord-developers, where all three actors collaborate to produce domicide. The demovicted Heron Gate residents are discursively positioned as not part of "our community," while the rest of Ottawa's majority white residents, as "team players," are expected to support their expulsion.

A subsequent slide presented at the open house event describes Vista Local as producing "liveable homes" and "safe and healthy communities" (see Figure 7-2). In this imaginary, the future of Heron Gate is liveable, the premium lives inhabiting these new homes hold value; all that is required to achieve this potentiality is the rendering of existing lives in Heron Gate as devoid of value in order to facilitate mass demoviction and domicide. Renditions of Vista Local in the planning and design stage indicate that "liveable homes" are only produced by the

Demoviction 2016: Domicide and Redevelopment in Heron Gate 89

Heron Gate: Now & Future

The community is improving:
- Heron Gate Mall - 2012
- Finn Court/Colbert Court - 2010
- Ledbury - 2008
- Heatherington/City Works Yard – Under Review
- Canada Lands

Heron Gate Renewal:
- Heron Gate phase 1 - 348 residences – 2017
- Future phases?

Heron Gate: 39.54 acres in Alta Vista

Timbercreek

- Purchased Heron Gate site in 2012
- Invested $40M in upgrades
- Investing $120M in a new building
- Future investment is best coordinated starting with a Secondary Plan
- This is a team effort:
  - Timbercreek
  - Team of international experts in community design
  - City of Ottawa
  - And you – our community

Timbercreek

DIALOG

Ottawa

Figure 7-1 Presentation slides prepared for an open house event outlining a preliminary sketch for "Heron Gate renewal," undertaken as part of a "team effort" between the landlord-developer and the city (City of Ottawa 2018).

inhabitancy of white people (see Figure 7-3), an implication not lost on organizers with the Herongate Tenant Coalition:

> This type of destructive behaviour of financialized landlords disproportionately targets people of colour…. You're dealing with like an ethnic historic Somali neighbourhood, where there is a highly valuable social fabric that exists amongst the population there. And you are replacing it with a completely different group of people that are middle- to upper-middle-class white people that don't know each other. And so it's a huge shift. And they made fucking paintings of that. (Anonymous interview, June 7, 2021)

90   RESISTING EVICTION

Heron Gate in the future: HG7

- Liveable homes
- Safe and healthy communities

Figure 7-2 Presentation slide prepared for an open house event promising "liveable homes" and "safe and healthy communities" for Heron Gate (City of Ottawa 2018).

Figure 7-3 Design rendition of the Vista Local redevelopment with no sign of the Black and Brown families living in Heron Gate (Timbercreek Asset Management 2016b).

One person I interviewed who still lives in the neighbourhood emphasized that those Black and Brown families remaining in the neighbourhood were highly aware not only that their community was being replaced but also that the gentrification and displacement happening in Heron Gate was part of wider and historic trends in Canada:

> Even in their visual graphics ... if you actually look at the little stick figures they used to kind of represent people and their little coffee shop in the development, like it's *all white people*,

and they're actively replacing the demographic here. And that's definitely interesting to me. It definitely speaks to, I think, a lot of what's been happening not only in Ottawa but across Turtle Island, like what happens in Jane and Finch and Regent Park and in Parkdale, in all of these places like Africville. It is not lost on us.... People here are very aware of this. (Anonymous interview, May 27, 2021)

The same interviewee spoke of another image displayed at one of the open house meetings organized by Timbercreek and the City of Ottawa:

> There was a graphic that had been going around talking about how wouldn't it be so clean and lovely to have a little coffee shop here, completely disregarding the fact that there were families and children who had lived here. But, you know, it's fine for wealthy people to have a coffee shop in place of that.

References in developer visualizations to "retail" as some vague promise of consumer trinkets, as well as to coffee shops and leisure, speak to the settler desire for consumable spaces in pristine settings, where property is neat and tidy and welcoming to a white, affluent lifestyle, despite the real human costs of what it has replaced. White enjoyment of Heron Gate is only realizable through the erasure of existing residents and homes, just as settler advantage and wealth is only attainable through the theft of Indigenous land.

Urban revanchism and elimination of the "mess"

The disdain held by some Alta Vista property owners for lower-income Heron Gate tenants, especially Black residents, is captured in documents obtained from the City of Ottawa through FOI requests. One such document is a report from Suzanne Valiquet of Momentum Planning & Communications on the September 2016 public consultation event, where Timbercreek proposed resort-style rental accommodations for Heron Gate (City of Ottawa 2020-424). Valiquet has worked for Timbercreek as a public relations consultant tasked with "overseeing the Communications and Relocation Program at Timbercreek's Heron Gate Development," according to her LinkedIn page (Rockwell 2021, 79–80). In the Momentum report, Valiquet documents in detail conversations that were held largely between Timbercreek's Greg Rogers, planning consultant Jack Stirling, and a group primarily made up of Alta Vista homeowners. One participant asked, "What are you going to do with the rest of the mess (garden homes)?" Another question began with a statement: "This is a ghetto. How do you make sure those people don't move into this new place? Even in smaller apartments, two Somali families can live in one unit?" Rogers ignored the racist aspects of the comment and simply replied: "The short answer is price — large units in an expensive building" (City of Ottawa 2020-424, 297).

This type of community vision and consultation forum on the Heron Gate redevelopment demonstrate what Neil Smith (1996) identifies as a form of revanchist attitudes toward lower-income and racialized residents. Revanchist politics are seen in the desire of non–Heron Gate residents to remove racialized tenants from the neighbourhood. According to August (2014b, 1,162), "Rather than a benign or natural process, revanchist politics reveal how gentrification is a product of political and cultural struggle in which elite groups seeking to defend race and class privilege assert their entitlement to desirable areas and valuable real estate." The revanchism of Alta Vista residents represents a desire to not only defend lines of white privilege, what Smith (1997, 122, cited in August 2014b) refers to as a defensive "rehardening of white middle class identity," but also spread their socioeconomic and racial identity beyond their own neighbourhood in order to destroy and remake Heron Gate.

The Momentum report documents Rogers reiterating the purpose of redeveloping Heron Gate to align with Alta Vista. During the meeting, Rogers stressed to Alta Vista residents that his "goal is to convince you that you want to live there" (City of Ottawa 2020-424, 300). When questioned about whether the new development would contain affordable family units, Rogers replied that "large premium units ... will be offered at a premium price. They will not meet affordability housing requirements in any way" (City of Ottawa 2020-424, 298). The absence of affordable housing works to ensure that existing Heron Gate residents will have no home in the Vista Local redevelopment. Rogers also admitted that the redevelopment is not designed to accommodate the large, multigenerational, and predominantly Black and Brown families that live in the townhomes: "Townhomes were occupied by families as large as 10, etc. The building we want to build would be occupied by smaller families, young professionals, retirees, probably an average of 3 per unit." The Momentum report shows how Timbercreek's approach to creating a revitalized Heron Gate involves mass eviction, demolition, and the creation of conditions where the existing residents are unable to relocate within the neighbourhood.

One resident I interviewed for this research discussed the impacts of the broader effort to exclude Somali families from a redeveloped Heron Gate:

> So there's very much a feeling of like, honestly, almost collective trauma from seeing everyone that you love and know in this area who you've grown up with, knowing now that they are in a precarious housing situation, that they've lost their community, that our parks are empty, that there are no more kids hanging out, having a good time growing up, seeing people who look like them, you know, knowing that kids as Black and Brown kids, especially in this area, no longer feel safe, and they feel overpoliced. They feel policed in general, they feel uncomfortable, seeing gentrified units come up. It's just, it's not the same area we knew and that we grew up in. (Anonymous interview, May 27, 2021)

This resident spoke about the sense of insecurity plaguing their family and other remaining residents, as they could receive an eviction notice at any time giving them three months to leave. They also spoke about the emotional effect of watching their neighbourhood being dismantled in terms of Black and Brown residents being replaced with new white residents in the new development, where the new residents treat the remaining racialized tenants with suspicion:

> I remember a few weeks ago I was walking late one evening and I have to walk past the area that has been demolished and replaced with luxury condos. So I have to walk past there at night. And it's so funny because I always feel a very real sense of like being very unnerved walking past that area at night. And it's funny to me because there's also a sense of rage, because I frequently see security patrolling around this new development. And it's like insane to me that this company will force out Black and Brown families, displace them and then replace them with wealthy white people, and then have the audacity to hire security to save these white people from the Black and Brown people left in the area.

This is a powerful observation from one remaining resident gazing upon the new structure. The gaze, however, is returned from those now living in Vista Local, and it is not necessarily a friendly one:

> It's very funny because I get dirty looks now when I walk past it and I'm like, "Okay, all right. I've lived in this area for a while now, you can relax. Like I'm not doing anything to make you feel uncomfortable.... I have no sense of feeling any type of rage at you." But then I do often feel slightly resentful that, you know, my home and my community is being destroyed so that people can come and live here and feel, you know, comfortable in this area. But again, what is the human cost of this, of your comfort? So I think about that sometimes.

This experience demonstrates the revanchist attitudes that have accompanied the gentrification of Heron Gate. In this example, white residents who have moved into the new Vista Local redevelopment seek to assert their entitlement and defend their race and class privilege by attempting to make long-time residents feel unwelcome in their neighbourhood.

The revitalization of Heron Gate is producing "liveable homes" on the remains of the affordable dwellings of an ethnoracial enclave. To justify demoviction and domicide, the existing homes and community had to be rendered unliveable, the existing residents embodying unliveable life, life that has less value, life that does not count. This is what domicide looks like under practices and processes of financialized gentrification in a white settler society. In the case of Heron Gate, settler colonial logics of racialized property relations and revanchist urbanism are inscribed with dominant (white, settler) views that lower-income and racialized subjects (tenants) are incapable of increasing land values, and are therefore preventing desirable subjects (property owners) from extracting the highest and best use (the exchange value) from the land. The attempted obliteration of an ethnoracial enclave through mass demolition-driven eviction, in order to build "premium units ... at a premium price" for Alta Vista demographics, is the subject of a human rights legal challenge that is discussed in more detail in the concluding chapter.

An investigation into the Vista Local redevelopment has provided a window into the various approaches and strategies that financialized real estate firms deploy in the built environment in order to attempt to perfect property relations over the longer term and build more wealth for investors. Representing a twenty-first-century version of urban renewal efforts under the benevolent guise of revitalization, demoviction is a distinct tactic deployed within the broader contemporary trend of financialized housing. While gentrification is produced through discourses of improvement — through mechanisms of intensification, revitalization, and liveability — and enacted through financial mechanisms, domicide is neither inevitable nor uncontested. The next chapter examines tenant resistance to financialized gentrification and how landlords respond in turn. Resistance at Heron Gate is ongoing and has played out in social media, in the streets, and in the courts.

CHAPTER 8

Demoviction 2018: Tenant Resistance to Domicide

The first phase of demoviction and redevelopment at Heron Gate was just that, only the first. Though not necessarily evident at the time, Timbercreek had launched a plan to demolish the remaining parcels of rental townhouses. One clue was that in describing the new development as "seed[ing] change" throughout the area (McCracken 2016a), Timbercreek's Greg Rogers stated that the company was looking at similar options elsewhere in the neighbourhood (Shaw 2017). Having faced little opposition to the first round of mass evictions, an emboldened Timbercreek was set to announce plans for an even larger demoviction project.

On May 7, 2018, Heron Gate residents were called to a "resident information session" with Timbercreek officials and Councillor Cloutier, where they each received an eviction package. A letter in the package implicated over one hundred families living in low-rent townhouses between the borders of Heron Road, Baycrest Drive, and Sandalwood Drive. In the letter, property manager Paul Boutros declared that the homes were "reaching the end of their building life cycle" and that 25 percent of these were "no longer viable" (Timbercreek Communities 2018a).

As in 2016, Timbercreek officials justified the demoviction by claiming it was the only viable option. They gave residents a deadline of September 30 to vacate. While the first demoviction in 2016 had angered tenants and the wider community, the announcement of the second set of mass evictions in 2018 prompted residents to organize and fight back.

Unwilling Subjects of Financialized Gentrification

Research has examined the dynamics of contention and negotiation between developers and social actors regarding housing policy in urban environments (Chisholm, Howden-Chapman, and Fougere 2020; Domaradzka 2019), including developers' evolving strategies to establish and maintain legitimacy over the production of space (Hyde 2022; Mosselson 2020; Robin 2018). With the growing creep of finance capitalism into the social sphere and housing sector, insecurity and inequality become heightened (August and Walks 2018; Fields and Uffer 2016; Soederberg 2018, 2021). Community organizations and other social actors have engaged in an oppositional politics to contest the adverse effects wrought by the financialization of the housing sector, chief among them displacement (Fields 2015, 2017; Teresa 2019).

Of course, tenant organizing has not only emerged in response to the financialization of rental housing and to the corporate capture of apartments by large real estate investment firms. Tranjan (2023) documents the recent history of tenant organizing in cities across Canada, and tenant organizing has a long history on a global scale. Tenant movements in Ontario in recent years provide an appropriate starting point from which to further engage with the struggle at Heron Gate and the mobilization of the Herongate Tenant Coalition. August and Webber (2019) document the emergence of grassroots groups of tenants organizing around housing justice issues in their Ontario communities. The authors outline some of the best practices for grassroots community organizing, including directly targeting adversaries, linking struggles, remaining independent, and organizing at the local level. They refer to this last point as "district-based scale," while Webber and Doherty (2021) refer to it as "territorial organizing."

This type of organizing is targeted toward and situated within the neighbourhood or building level. The emphasis is on localized struggle that mobilizes directly impacted residents while not succumbing to larger social and political forces (such as outside groups, nonprofits, and other organizations with goals and mandates not directly related to the struggle). Parkdale Organize, which embraces this model and organizes on a variety of fronts, including tenant and housing justice, has been more militant in its approach than other groups. It has organized successful rent strikes (Shilton 2021) and inspired other movements such as the Hamilton Tenants Solidarity Network to attempt to organize rent strikes of their own against landlord InterRent REIT (Power and Risager 2019; Risager 2021).

The Herongate Tenant Coalition has worked in the spirit of directly targeting its adversaries while focusing on organizing residents at the local level. At the same time, the coalition has successfully sought out a range of allies to assist with other aspects of the struggle (e.g., media, fundraising, and legal). It has launched fierce and effective media campaigns, including mobilizing social media to publicly shame adversaries, working with independent media to publish stories about the struggle, and garnering significant local, national, and international mainstream media attention on the mass evictions.

Analysis of the Herongate Tenant Coalition's actions against Timbercreek helps build on Fields' (2017) theorizing of tenants as "unwilling subjects of financialization," further challenging the notion that the financialization of rental housing is an uncontested inevitability. In this chapter I articulate the tactics deployed by the Herongate Tenant Coalition to try and stop the evictions as well as to challenge the landlord, underlining the important theorization and knowledge production work in which it has necessarily engaged as part of this struggle. I also highlight the tactics deployed by apartment investors in response to this opposition, extending the documentation of landlord efforts to demobilize tenant resistance (August 2016). Thus, this chapter documents both tenant organizing against ruling institutions and examines how ruling institutions respond to tenant resistance. In the Heron Gate case, the threat of mass eviction catalyzed tenant resistance.

Mobilizing against mass eviction

Shortly after Timbercreek issued the May 2018 eviction notices, the Herongate Tenant Coalition (2018b) formed to challenge and resist what it described as the "largest eviction and displacement campaign in Canada." Tenants from the Herongate and Heatherington neighbourhoods made up the bulk of coalition members. The coalition quickly pulled together a tenant rights meeting, held in the community on June 2, which was attended by dozens of residents and facilitated by coalition member Ikram Dahir, who also provided interpretation services in Somali (Herongate Tenant Coalition 2018c). Also present at the meeting to field questions about tenants' legal rights under domestic and international law were lawyer Daniel Tucker-Simmons and Leilani Farha, a Canadian appointed as the United Nations special rapporteur on adequate housing. That meeting was a galvanizing moment when dozens of Heron Gate tenants, in particular a significant number of racialized women, joined the coalition and embraced leadership roles.

According to its website, the coalition is organized and run by the working-class people of Herongate, and "was formed to build power, strength and solidarity … in this moment of crisis so we can have each other's backs, care for each other and defend our neighbourhood from development, speculative and political forces that want us out of here" (Herongate Tenant Coalition 2018d). One organizer described the group's aim to "formalize a tenant-based neighbourhood organization specifically of tenants, and to make it mainly associated with people that actually live and rent in the neighbourhood, as opposed to it being sort of a pet project of city progressives" (Anonymous interview, June 7, 2021). While other more established community organizations are attempting to influence the landlord through negotiation, the Herongate Tenant Coalition, as a small collective with no formal structure, has adopted a radical analysis and praxis of refusal, mobilizing tenants to refuse eviction — to stay put.

One coalition strategy was to actively encourage tenants to defy Timbercreek's attempts to get them to end their tenancy. Timbercreek's efforts to remove tenants centred on convincing each household to sign an N11: Agreement to End Tenancy form, in essence voluntarily agreeing to end their tenancy. If tenants did so, Timbercreek could avoid having to seek an eviction order from the Landlord and Tenant Board. The Herongate Tenant Coalition launched an information campaign, which included materials translated into French and Somali, to notify tenants of their rights and let them know they were not legally obligated to sign the N11 form (see Figure 8-1).

This campaign represented a significant threat to Timbercreek's authority and control over the neighbourhood, its objective of reconfiguring the community in line with the more affluent Alta Vista, and, relatedly, returns on investment. In response, Timbercreek issued a series of letters to tenants, threatening as "a last resort" to apply to the Landlord and Tenant Board for a forced removal if tenants chose to remain in their homes after the September 30 deadline. Such action, though, would create an ugly political scene that would no doubt further damage the landlord's reputation.

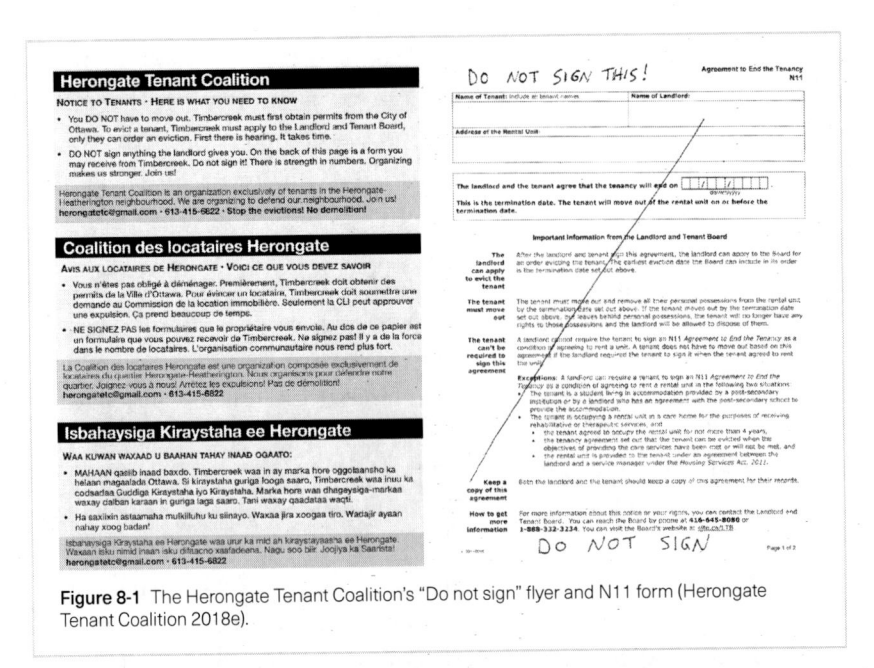

Figure 8-1 The Herongate Tenant Coalition's "Do not sign" flyer and N11 form (Herongate Tenant Coalition 2018e).

Then a mysterious poster appeared throughout the neighbourhood. Containing no logo or signature, it urged tenants to "Know Your Rights" in a font that looked more like the work of an activist group than a corporate or financialized landlord. The poster empathized with the soon-to-be displaced, stating: "You deserve honest advice from someone who will tell you and your neighbours the truth" and "Get the truth, for you and for your family" (see Figure 8-2). The poster, presumably designed by people working for Timbercreek or Momentum, does not look like something created or sanctioned by the landlord; however, it encourages tenants to "make the right call about having to move." Coalition members viewed this tactic as part of a wider disinformation campaign meant to sow doubt and confusion about the implications of refusing to move while painting them as illegitimate outside agitators trying to implement a malicious agenda on genuine Heron Gate tenants.

To counter the "mass eviction" narrative of the Herongate Tenant Coalition, Timbercreek framed its operation as a "relocation program" in various forms of communication, including posters. As with the 2016 demoviction, Momentum Planning & Communications was tasked with facilitating the removal of residents from their homes and having them find alternative housing. However, whereas in 2016 some displaced families were allowed to move within the neighbourhood, often into vacant

> # KNOW YOUR RIGHTS
>
> You deserve honest advice from someone who will tell you and your neighbours the truth.
>
> ## Make the right call about having to move
>
> - *Landlord and Tenant Board* **1-888-332-3234**
> - *Law Society of Ontario Referral Service* **1-855-947-5255**
> - *Pro Bono Ontario* **1-855-255-7256**
> - *Legal Aid Ontario* **1-800-668-8258**
>
> ## Get the truth, for you and for your family

Figure 8-2 "Know Your Rights" poster, presumably circulated by agents of Timbercreek or Momentum Planning & Communications (Timbercreek Communities 2018b).

townhouses in considerable disrepair, in 2018 Timbercreek attempted to relocate households outside of the community. Initial landlord communications about the relocation program indicated the option of being able to move elsewhere within the landlord's Heron Gate holdings; however, a relocation drop-in session notice then stated that no homes were available in Heron Gate and that Timbercreek would assist tenants in finding accommodations outside the community. The notice encouraged residents to move out as soon as possible, warning that fewer rental units would be available in the city at the end of the summer. The Herongate Tenant Coalition insists the landlord undertook such measures to generate fear and hasten the removal of tenants.

The demoviction survey

The Herongate Tenant Coalition undertook a door-to-door survey of the implicated parcel of townhomes in spring 2018 to document the neighbourhood demographics and understand who exactly would be impacted by the evictions. It documented the addresses in a map on its website, and compiled findings into an infographic (see Figure 8-3). While Heron Gate is large enough to have its own census tract, the coalition wanted more precise data as part of its research and mobilizing efforts. The survey revealed that 93 percent of the impacted households were racialized, including 49 percent of Somali descent (Herongate Tenant Coalition 2018a, 2018b). Almost six hundred people were recorded by the survey, including more than two hundred children. The survey also captured the gendered dynamics at play, as single mothers led many of the households (2018a). The families were also large, with an average of 5.4 people per household.

In addition, the Herongate Tenant Coalition compiled and organized data on poverty, housing, and immigration obtained from the 2016 Statistics Canada Census to compare quality of life and social indicators in Heron Gate with those of the largely white, affluent Alta Vista neighbourhood as well as Ottawa-Gatineau as a whole. The results, which further supplement the data presented in Chapter 5, demonstrate the neighbourhood's high rates of low-income residents, children

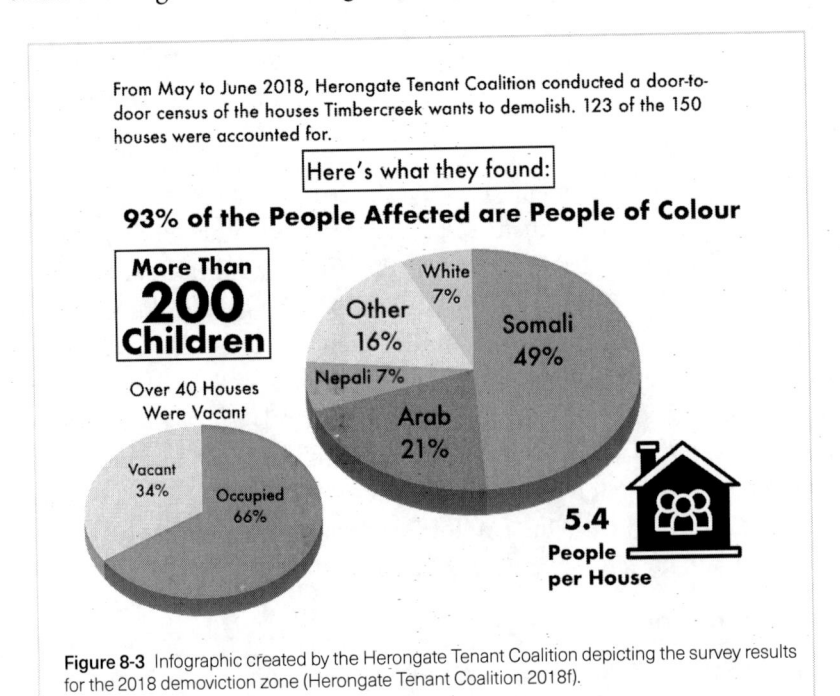

From May to June 2018, Herongate Tenant Coalition conducted a door-to-door census of the houses Timbercreek wants to demolish. 123 of the 150 houses were accounted for.

Here's what they found:

93% of the People Affected are People of Colour

More Than **200 Children**

Over 40 Houses Were Vacant

White 7%
Other 16%
Nepali 7%
Arab 21%
Somali 49%

Vacant 34%
Occupied 66%

5.4 People per House

Figure 8-3 Infographic created by the Herongate Tenant Coalition depicting the survey results for the 2018 demoviction zone (Herongate Tenant Coalition 2018f).

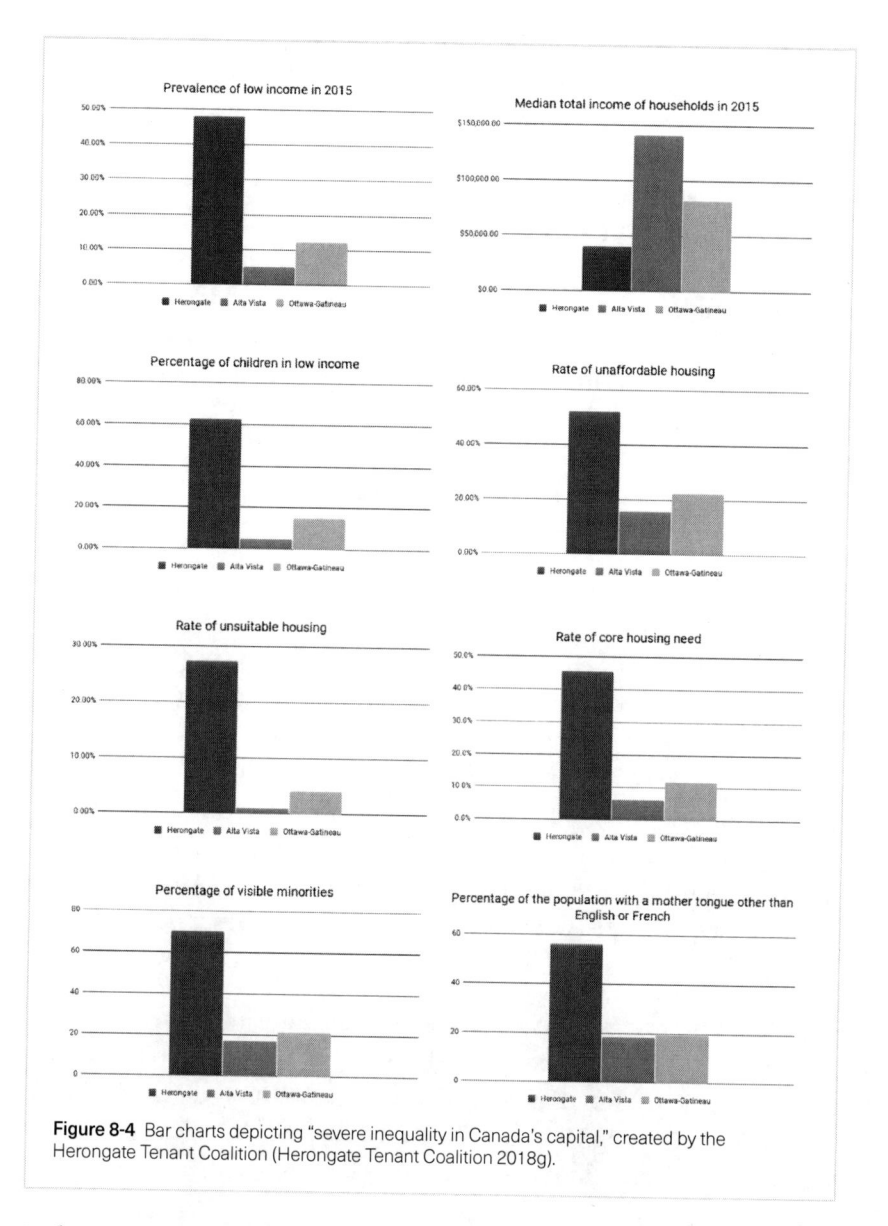

Figure 8-4 Bar charts depicting "severe inequality in Canada's capital," created by the Herongate Tenant Coalition (Herongate Tenant Coalition 2018g).

in low-income homes, core housing need, visible minorities (around 70 percent compared to Ottawa-Gatineau's 20 percent), and residents whose mother tongue is not English or French (Canada's two official languages) (Herongate Tenant Coalition 2018g). The coalition visualized these aspects of structural inequality and discrimination, where poverty intersects with racism and housing, in a variety of charts (see Figure 8-4).

The prevalence of low-income households in Heron Gate was almost 50 percent in contrast to just over 10 percent in Ottawa-Gatineau and around 5 percent in Alta Vista. Likewise, the percentage of children in low-income households, at over 60 percent, is staggering compared to the Alta Vista and wider regional figures. The high rates of visible minorities and residents speaking a first language other than English or French are reflective of Heron Gate as an area containing many immigrant and refugee families. Housing statistics in relation to core housing need, which were explored earlier in Chapter 5, are striking when juxtaposed alongside levels in Alta Vista and Ottawa-Gatineau. Perhaps most significantly, the median total income of Heron Gate households is less than half of the Ottawa-Gatineau average and less than one-third of the average in Alta Vista. The Herongate Tenant Coalition (2018g) notes the significant inequality captured here:

> The links between housing, inequality and structural racism are very present in Herongate. In fact, nowhere in Ottawa are these issues tied so deeply together. While every level of government and Timbercreek are determined to destroy our neighbourhood, a neighbourhood that should in fact be treasured, celebrated and supported, they fail to realize the organizational capacity we possess and our own determination in fighting to save our neighbourhood. Herongate sits across the street from one of Ottawa's wealthiest areas – Alta Vista. The inequality between these two neighbourhoods is huge. (Herongate Tenant Coalition 2018g)

This stark contrast is crucial for contextualizing Timbercreek's efforts to align the neighbourhood with Alta Vista as well as the relevance of affordability metrics in the wider redevelopment proposal (examined in Chapter 9), as existing Heron Gate residents will not be able to afford the rents earmarked as "affordable" in the new development.

Public Shaming, Social Media, and Legal Repression

The Herongate Tenant Coalition has engaged in a variety of tactics with the ultimate aim of organizing tenants to refuse to move. To that end, it coordinated community meetings, demonstrations, neighbourhood walking tours, and submissions of maintenance work orders by tenants to pressure the landlord to repair their homes. In tandem with these efforts, the coalition unleashed an unrelenting social media campaign that directly targeted Timbercreek and City of Ottawa personnel for their roles and responsibilities in mass evicting and displacing a largely racialized community. This social media campaign constituted part of a larger strategic repertoire used by social movements known as public shaming, which "occurs when large numbers of people engage in the collective shaming of some individual or organisation perceived to have done something morally wrong" (Fox 2020, 177). While the targets of public shaming can find it deeply distressing and degrading (Fox 2020), the tactic can be an effective tool for grassroots movements with limited resources struggling against powerful actors (Mielczarek 2018; Vanderheiden 2021).

Tenant organizers have embraced the tactic of public shaming to expose the abuses of landlords to the wider public. This can happen through a poster campaign, demonstration at the home of a landlord, or social media. In the case of the Heron Gate mass evictions, the Herongate Tenant Coalition used social media to profile Timbercreek executives and employees, their involvement in the evictions, and their ties to other actors, including city officials. Instead of responding with another alternative information campaign, such as the "Know Your Rights" poster described above, Timbercreek invoked legal mechanisms to address the public shaming.

Landlord: "Cease and desist"

Over the course of four months, Timbercreek lawyers issued a series of cease and desist notices to the Herongate Tenant Coalition. The coalition's social media campaign had drawn a wide range of support, sympathy, and media coverage by calling attention to the practices of Timbercreek executives and their connections with city officials. Timbercreek's campaign of legal threats and intimidation in an effort to demobilize the Herongate Tenant Coalition's social media campaign provides insight into how landlords may respond to tenant organizing through tactics of repression.

The first letter, dated July 9, threatened legal action, claiming damages to Timbercreek and its employees resulting from six social media posts, including posts made by non-coalition members who merely tagged the coalition's Twitter handle. Timbercreek's lawyer claimed the posts were defamatory, in violation of the *Libel and Slander Act*, and ordered the posts removed immediately. Coalition members perceived the letter as a scare tactic with the primary aim of intimidating people away from organizing and speaking out against the evictions. Refusing to concede to Timbercreek's threats, the coalition continued to post regularly to social media. Persistent, Timbercreek's lawyer followed up with similar letters dated August 7, August 30, September 6, and October 5, outlining alleged defamatory social media posts and demanding the coalition "cease and desist."

The coalition published the cease and desist letters online, an action that Timbercreek took considerable issue with. In a letter to Timbercreek's lawyer, Herongate Tenant Coalition lawyer Daniel Tucker-Simmons asserted the coalition's legal right to share the letters as it saw fit, explaining the coalition's perspective that transparency with the larger community was important for drawing attention to Timbercreek's intimidation tactics:

> Though you are unfamiliar and uncomfortable with activism, and likely consider it "unprincipled" in comparison to principled discussions between lawyers, you'll appreciate that my clients and their allies recognize that their limited resources put them at a great disadvantage in legal proceedings. For that reason, they prefer to engage your client in an arena in which the playing field is somewhat more level. (Tucker-Simmons 2018)

I asked Tucker-Simmons to elaborate on some of these points in an interview (October 13, 2018). He noted that corporations are typically not agile at engaging in the activist arena and prefer to use legal resources and methods as a multipronged strategic approach. Beyond the immediate objective to suppress and silence criticism, they have a secondary objective to drain resources away from the activist campaign. Tucker-Simmons explained that when activist campaigns or public interest organizations are forced to engage well-resourced corporate entities in the legal arena, it has the threefold effect of draining financial, human, and emotional resources. And that is precisely the political intention of corporations engaging in legal tactics against social movements: to divert resources and energy away from the campaign to the court, to divert adversaries from their activism and into passivity and docility

(Landry 2014). The threat of a lawsuit is a tactic on its own. Through legal threats, corporations attempt to gain the upper hand by baiting or compelling an opposing campaign or movement into the legal arena. The Herongate Tenant Coalition's defiance was an attempt to level the playing field against a powerful economic actor.

Legal repression: SLAPPing social movements

To make sense of Timbercreek's cease and desist threats against the Herongate Tenant Coalition, here I explore the demobilizing tactic of strategic lawsuits against public participation, commonly known as SLAPP suits (Beder 1995; Canan 1989; Hilson 2016; Landry 2014; Sheldrick 2014). Discussions of political repression tend to focus on the state and protest control (Boykoff 2007; Davenport 2007; Earl 2011), but legal repression can include nonstate parties who seek to use the law to dismantle social movements (Ellefsen 2016). Legal repression has been defined as "a process whereby the state and/or non-state elites attempt to diminish dissident action, collective organization, and the mobilization of dissenting opinion by inhibiting collective action through raising the costs and/or minimizing the benefits of such action, by way of law and criminal justice" (Ellefsen 2016, 445). Demobilizing tactics that move beyond state actors and protest control to incorporate the threat of sanction under civil law are a creative form of repression within liberal democracies.

In the realm of social movement suppression, SLAPPs are a tool of intimidation used to silence opposition. They can involve an actual lawsuit or just the threat of a lawsuit (White 2005). The SLAPP connotes a sudden violent affront, intended to shock, as a metaphorical slap in the face of the targeted victim (Landry 2014), but also represents a "SLAPP in the face of democracy," as Donson (2000) suggests. Landry (2014, 7) describes SLAPPs as a "deliberate instrumentalization of legal proceedings as a weapon of intimidation, censorship and political reprisal in social and political conflicts."

SLAPPs tend to be deployed by capital/corporate entities against opposing social actors, with the goal of stifling dissent and negative publicity. To this end, SLAPPs often rely on allegations of defamation, the communication of a false statement that inflicts reputative harm. For White (2005, 272), SLAPPs are deployed "not to 'win' in the conventional legal sense, but to intimidate those who might be critical of existing or proposed developments." The corporate arsenal increasingly features SLAPPs as associated costs are rationalized as a minor part of the cost of doing business, whereas for individuals or groups a court case could result in bankruptcy (White 2005). Pring and Canan's (1996, xi–xii) research examining thousands of cases in the United States shows that (1) the legal system is ineffective at controlling SLAPPs; (2) SLAPPs profoundly affect political outcomes; (3) SLAPPs are seldom won in court yet achieve political goals; (4) SLAPP defendants rarely lose in court yet are "frequently devastated and depoliticized"; and (5) SLAPPs institute a chilling effect where people are discouraged from speaking out.

Real estate development SLAPPs represent the single largest category of strategic lawsuits in the United States, comprising over one-third of all cases examined by Pring and Canan (1996). While the legal and constitutional landscape differs considerably in Canada (see Landry 2014; Sheldrick 2014), the experience in the United States can provide insight into the motivations of corporate developers and financialized landlords in Canada. According to Pring and Canan, real estate SLAPPers can be motivated by a variety of reasons — both material and value-laden — including obligations to investors, commitment and contribution to economic growth, and the desire to sanction opponents perceived to be acting in bad faith. One U.S. attorney who has represented both sides of real estate development SLAPPs maintains that SLAPP suits are "viable weapons" when the SLAPPer has a good reputation, the profit stakes are high, the opponent engages in personal attacks, and "the targets are relatively unsophisticated individuals" lacking institutional support (Pring and Canan 1996, 42).

It is interesting to consider these circumstances in the case at hand. First, in the lead-up to the 2018 mass eviction notice, Timbercreek had a relatively unsullied reputation; however, the firm rebranded shortly thereafter in 2020. The damage to its reputation as a result of coalition activism could partly explain why Timbercreek did not follow through with legal action. Second, the profit stakes are incredibly high, as Heron Gate represents a valuable asset and investment with billions of dollars tied into the redevelopment. Third, the Herongate Tenant Coalition has engaged in public shaming of Timbercreek executives — typically white, wealthy men — to personalize the struggle, showing the human beings behind decisions to remove racialized people from their community. Finally, the last point on targeting "relatively unsophisticated individuals" through SLAPPs may help explain why Timbercreek did not follow through with legal action. Despite engaging in unconventional tactics, the Herongate Tenant Coalition has led an incredibly sophisticated, well-researched, and principled campaign to inform tenants of their rights under landlord-tenant law as well as to attempt to keep people in their homes. For its efforts, the coalition has been vilified by the landlord, city officials, and even local civil society leaders.

In the end, the lawsuit never materialized and Timbercreek was successful in removing more than a hundred Heron Gate families from their homes in 2018. The coalition's mobilization efforts continued, however. A neighbourhood protest would become a flashpoint in the next stage of the fight between the Herongate Tenant Coalition and Timbercreek.

The Timbercreek Twitter affair

Under the threats levied by Timbercreek employees, tenants felt it was too risky to stay and defy the eviction order. By the first week of October 2018, most of the evicted families had moved out of the neighbourhood. To coincide with the mass displacement, the Herongate Tenant Coalition organized a rally and walking tour of the neighbourhood on October 4. The rally drew support from outside the community as well as local media. Participants toured the mainly empty streets, showing the derelict buildings to neighbourhood outsiders bearing witness to the systemic neglect that pushed these homes to the "end of their life cycle." The walking tour concluded at Timbercreek's neighbourhood office. Despite the peaceful nature of the rally and its relatively small size, employees locked themselves inside and called the police. The involvement of Ottawa police became a central element in what unfolded next.

Early in the fall, the landlord had started to apply more pressure on the coalition's online activities. On September 25 the coalition received an email from Twitter's legal team that included a letter from Timbercreek's law firm. Timbercreek had asked Twitter to remove three of the coalition's tweets. Twitter specified that it had not taken any action with regards to the coalition's account.

Although most evicted people had moved by the September 30 deadline, Timbercreek was still invested in silencing public criticism of its record. On October 11, Timbercreek's lawyer sent another letter to Twitter, using the rally to fabricate a narrative of criminality and extremism and request that Twitter "permanently disable" the coalition's account. The letter claimed that the coalition's Twitter account was used to organize an "escalating campaign of harassment" culminating in the October 4 "incident" where a group of thirty people "attempted to force their way into Timbercreek's offices." The letter also claimed that five Timbercreek employees were "trapped" and forced to leave under police escort and that two coalition members were arrested and charged with trespassing. However, no participants in the rally had attempted to forcibly enter the office, Timbercreek employees did not exit under police escort, and no arrests or charges occurred.

The letter continued:

> HTC is responsible for an ongoing and escalating campaign of harassment and intimidation against the employees of Timbercreek, which has already required police intervention.... Ottawa authorities have described to us the principals responsible for @herongatetc as "unstable", "unhinged" and "extremist" and have warned our clients of the likelihood of further escalating activity. Should Twitter fail to permanently disable @herongatetc, it will be responsible for facilitating a campaign that has already resulted in criminal behaviour and is likely to continue resulting in such criminal acts.

Timbercreek's claim that coalition organizers were "unstable," "unhinged," and "extremist" is a crass appeal to racialized discourses surrounding criminality and terrorism, as well as to discourses of violence, given that mental health, crime, and extremism have racialized components. The "extremist" label is particularly revealing given its blurring with broader narratives of the war on terror (Onursal and Kirkpatrick 2021). "Being maligned as terrorists" by Timbercreek was particularly troubling for one resident and coalition organizer: "They knew at the time of 2018 when we were going around and doing the work that the majority of them were Black women, Black, Muslim, visibly Muslim women as well, and they felt comfortable literally calling us terrorists" (Anonymous interview, May 27, 2021). Timbercreek stigmatized its opponents as faulty subjects, as failed citizens incapable of embodying good, desired Canadian values (Nagra 2017; Thobani 2007).

Timbercreek's efforts to persuade Twitter to permanently disable the coalition's account continued. On October 25 Twitter's legal office sent another email to the coalition stating it had received a court order claiming that four referenced tweets were illegal. No details about the court order were provided. Twitter specified that it may be obligated to take action but for the time being was merely inquiring if the coalition would voluntarily remove the tweets. Having adopted a pragmatic strategy of refusal when faced with threats, the coalition has for the most part remained uncompliant and confident in maintaining the moral, legal, and ethical upper hand.

The letters to Twitter reveal how landlords are willing to marginalize opponents as well as how financialized firms respond to public demonstrations and unwanted media attention. The desperate nature of Timbercreek's efforts reflects the strength of grassroots organizing. The Herongate Tenant Coalition and other grassroots tenant movements have targeted landlords in a personal sense to humanize the ruling relations and resistance. Wealthy landlords and executives tend to present themselves as family men and benevolent contributors to the wider community (such as by playing golf for charity). Public shaming of ruling actors — high-powered real estate players and their political and legal backers — can be part of an effective political strategy of resistance and refusal that grassroots organizers can leverage in the absence of financial resources while signalling higher levels of risk to real estate investors.

Yet the coalition has also used other tactics in this struggle. In a surprising twist, the coalition has flipped the landlord's defamation threats.

Flipping Defamation: Suing the Landlord

The struggle between the Herongate Tenant Coalition and Timbercreek has been referred to as a David and Goliath type of struggle, where a group of tenants have taken on their multi-billion-dollar landlord, an entity armed with unlimited financial resources alongside the hubris to imagine that it could continue to evict hundreds of racialized people without facing opposition. Engaging in the legal system from a grassroots social movement perspective can be risky as well as politically fraught for more militant groups that are ideologically opposed to recognizing the legitimacy of ruling institutions and criminal justice actors. However, members of the Herongate Tenant Coalition decided to take the offensive against Timbercreek through legal action.

In May 2019 three members of the Herongate Tenant Coalition filed statements of claim with the small claims court. Two of the plaintiffs sought $25,000 each while the third plaintiff sought $20,000 in damages from Timbercreek. In response, Timbercreek filed a motion seeking to combine the three cases into one and another motion asking for the case to be dismissed and heard instead by Ontario's Superior Court. Attending hearings proved fruitful for gaining an understanding of Timbercreek's position and approach to the legal terrain. The firm's lawyers presented frivolous arguments and motions that were interpreted by the coalition's legal team as delay tactics meant to drag out the cases and increase the coalition's legal costs. Monetarily, these cases represented minuscule sums for Timbercreek — "nuisance money," according to the coalition's legal team. For the Herongate Tenant Coalition, the cases presented an opportunity to engage in the legal arena and learn more about its opponent and landlord tactics, as well as win small legal victories while minimizing legal costs in the event of losses at the small claims level.

The statements of claim articulate how Timbercreek attempted to suppress the organizing efforts of the Herongate Tenant Coalition. In one, the plaintiff claimed "the Defendant has engaged in a defamatory campaign to silence the lawful advocacy efforts by herself and HTC [Herongate Tenant Coalition] in order to protect its image and commercial interests." The plaintiffs give their version of events of October 4, 2018, and the rally in the community, which had triggered the Timbercreek Twitter affair. The plaintiffs asserted that they did not impede or disrupt the activities of Timbercreek employees and that Ottawa police officers were called and arrived after most rally participants had dispersed. Interactions with Ottawa police officers were limited to a brief conversation; there were no arrests, charges, or further interactions with police following that day.

One of the benefits of engaging the landlord in the legal arena is the ability to request documents relevant to the case, which the court can order parties to produce. The production of court documents in this case has supplemented the Herongate Tenant Coalition's research in identifying other players involved in efforts to criminalize and demobilize tenant resistance. Disclosure packages obtained from Timbercreek help fill in missing information and highlight some of the hitherto unknown players involved. Take, for instance, Suzanne Valiquet, Timbercreek's public relations consultant and overseer of tenant relocation in Heron Gate, who has served multiple terms on the Ottawa Police Services Board. The disclosure documents spotlight her relationship with Ottawa police and Timbercreek. On October 4, 2018, a 911 call was made from

Timbercreek's office requesting police assistance to disperse the rally. (We had earlier learned of this call through an FOI request.) Emails written by Valiquet to Timbercreek officials indicate that she received advice directly from the chief of police on how to escalate the police response: "At 5:39 pm I called 911. I called a second time (on the advice of the Chief of Police). This way our call went from a priority 2 to priority 1 emergency call. Six police officers arrived around 6:03 pm." Valiquet further indicated that the police officers advised her to continue to call for assistance if there were future demonstrations, noting that "the more times you call the bigger the file against [them] will grow."

The email is a damning indictment of police efforts to criminalize activists and landlord tactics deployed to criminalize tenant opposition, where a Timbercreek consultant may have leveraged her professional connections with the chief of police to obtain advice on how to intensify a police response to a peaceful demonstration. Valiquet's performance from within the Timbercreek rental office demonstrates a layer of interinstitutional synergy and strategy, in this case between a large developer and the Ottawa Police Services Board. One of the coalition plaintiffs made this observation in an email to the legal team: "The advice from the Chief was extremely reckless and an abuse of the 911 system. Escalating a non-violent, orderly protest and constitutionally protected form of free speech to a life-threatening emergency could have put Black lives in danger."

The small claims cases were settled out of court in 2022, and coalition members redirected their efforts toward building the legal defence fund and fighting a bigger case at the Ontario Human Rights Tribunal.

Landlords: "How to Handle a Crisis"

Tenant rights organizations like the Herongate Tenant Coalition, however small and lacking in financial resources, have been effective at pressuring landlords and drawing attention to issues of injustice. Wary of the potentially negative impact of unfavourable media coverage, landlords in Ontario are strategizing how to deal with tenant organizing. The Federation of Rental-housing Providers of Ontario (FRPO) organized a seminar in Toronto in March 2019 called "How to Handle a Crisis: The Media, AGIs and Tenant Actions." The event was promoted as a way to "strengthen members' knowledge and skills to effectively deal and communicate with tenants, and to handle media in a crisis." Sessions included "Changing the Landlord Image," "Crisis Communications and the Media," and "Managing the AGI Process." Themes throughout the panels included "changing the perception of landlords," managing the "agi/rent strike experience" in the media, and responding to "rent strikes and emergencies." Landlords clearly recognize that as a result of engaging in aggressive investment strategies and facing tenant resistance, the "landlord image" is in increasing disrepair, despite the dedication of extensive resources to advertising and public relations.

The Herongate Tenant Coalition presented a serious challenge to Timbercreek's efforts in the neighbourhood. It challenged not only the "relocation" narrative trumpeted by the landlord but also the eviction itself, thereby creating uncertainty and potential volatility around Timbercreek's investments. Across various financial instruments and risk forecasts, Timbercreek entices would-be investors with predictable, stable, long-term high yields in a "low-volatility" environment, having "earned a reputation for providing conservatively managed, risk-averse investment(s)" (Timbercreek 2019c, 2019e). Timbercreek's use of language reveals the investment-driven logics driving its interpretation of tenant resistance. Its investment products are marketed as stable, predictable, and risk-averse, in contrast to its framing of tenant opposition as unstable, unhinged, criminal, and extremist.

Coalition organizing and resistance has led to a number of legal challenges while also causing the landlord to pivot. In the months following the 2018 demoviction and leading up to the completion of the Vista Local redevelopment, Timbercreek rebranded as Hazelview. Further, the landlord was forced to rethink its redevelopment strategy for the neighbourhood and include concessions in its redevelopment proposal that included affordable housing and no future displacement of existing tenants. Although Timbercreek/Hazelview personnel would likely never admit that these changes were due to tenant organizing, the Herongate Tenant Coalition's interventions have had a wide impact in a complex struggle against gentrification, eviction, and displacement. Although many coalition members were displaced from the neighbourhood in 2018, a core group continues to mobilize and oppose the redevelopment process and plans for the neighbourhood, which is the subject of the next chapter.

CHAPTER 9

Community Wellbeing: A Social Framework for Domicide

In 2019 Timbercreek unveiled its master plan for Heron Gate, making presentations at a variety of public events over the next couple of years until the final approval of the Official Plan Amendment in September 2021. Not surprisingly, the company proposed large-scale neighbourhood redevelopment, entailing the demolition of hundreds more homes and the construction of thousands of new units. It based the master plan for Heron Gate on the Community Wellbeing Framework, a blueprint developed by the Conference Board of Canada for designing and redesigning communities that emphasizes the role of liveability in generating a return on investment while productively shaping resident conduct. Timbercreek announced that Heron Gate will become Canada's first community to be modelled after the Community Wellbeing Framework, making it a test subject for this particular model of liveability (Link2Build Ontario 2021).

The Community Wellbeing Framework also formed the basis for what came to be known as the "social framework" or "social contract," which was negotiated as part of the landlord-developer's application to the City of Ottawa for the larger redevelopment. A variety of actors participated in negotiations over the social framework, which culminated in an MOU governing the redevelopment agreement between the landlord-developer and the City of Ottawa. While the Herongate Tenant Coalition refused to engage in a process that it perceived as legitimizing the ongoing dismantling of the community, many other actors intervened to challenge the social licence granted to the landlord-developer from the City of Ottawa to demolish another 559 homes. This chapter investigates the production of the discourse of liveability within the context of Heron Gate, particularly in association with the social framework of community wellbeing governing the redevelopment of the neighbourhood.

Manufacturing Consent for the Master Plan

Developer actions and agendas are shaped by particular localized and place-based dynamics (Mosselson 2020). While property developers are often presented as powerful actors capable of exercising dominion over land in a linear fashion, Mosselson (2020) argues that a sociospatial perspective is warranted to account for how particularized settings influence developer approaches to urban landscapes. In particular, Mosselson (2020, 278) demonstrates the "contingent, socially and spatially embedded nature of developers," who adopt a "spatial praxis that requires adaptability and adjusting dispositions and practices" depending on the environments in which they operate. Timbercreek's approach to redeveloping Heron Gate provides a rich case study to understand the praxis of developers who must respond to various contingents of social and political actors with unique requests, demands, levels of bargaining power, and pressure tactics.

In the aftermath of the 2018 demoviction, Timbercreek and the City of Ottawa organized a number of meetings in the neighbourhood to showcase what they referred to as the "master plan" for redevelopment (Timbercreek Asset Management 2019g). They branded these events as "community visioning sessions" and "public open houses"; ward councillor Jean Cloutier, who consistently provided opening remarks at these meetings, called them a form of "preconsultation." In the final report from the City of Ottawa's (2021f) planning department recommending approval of the redevelopment plan, these sessions are referred to as formal consultations. Timbercreek and the City of Ottawa promoted these events as an avenue through which community members could directly influence the design of the neighbourhood. I attended all such events, the first of which was held in January 2019, right up until the final approval of the Official Plan Amendment in September 2021. At these events I collected audio data, took photos, talked to people, and recorded field notes. Throughout the process, similar sets of presentation slides were put on display, all sharing photos of previous sessions and emphasizing community engagement and the role of the public in providing input into the design process.

For both the Heron Gate redevelopment as well as the City of Ottawa's New Official Plan, public consultations have involved professionally certified designers and planners developing blueprints for redevelopment and liveability, and then engaging in performative acts of community engagement and consultation to seek buy-in. Research undertaken by August (2016, 29) demonstrates how community engagement sessions are actually designed to limit tenant interaction yet enable developers to market redevelopment projects as "tenant-driven." These processes, which simply inform residents of redevelopment plans, serve largely to manufacture consent and create the impression that community members themselves are informing and partnering in the design processes. A staff person for a city councillor put it this way in an interview: "We have this culture of consultation that usually produces results that are totally disconnected from what people actually said, and there are no teeth in that process" (Anonymous interview, June 18, 2021).

Open house events in February and March 2019 were held at the Heron Road Community Centre and organized similarly in terms of scope and layout. The main presentations were delivered in the community centre's third-floor theatre, which had a large screen and progressively inclined seating for roughly seventy-five people. A number of large print-out displays in the hallway featured the vision and principles guiding the redevelopment, as well as various design concepts. Timbercreek set up rooms where staff from urban design firm DIALOG exhibited three design concepts with the stated aim of soliciting feedback from attendees. I visited the three rooms, where Antonio Gómez-Palacio, an urban planner with DIALOG, and two other DIALOG staff fielded questions in relation to the different design concepts. The designs did not significantly differ from each other in terms of the layout and height of the buildings. The major differences seemed to coalesce around green space layout. I spoke to DIALOG staff about how the final design would be determined and whether these "consultations" were definitive. They told me that input from community members would be considered alongside that of the primary decision-makers, including city officials, planners, and Timbercreek staff.

The open house events followed a similar format. Cloutier would deliver some opening remarks, and then Timbercreek's Greg Rogers would make a presentation repeating the landlord's public relations lines, that Timbercreek approached the Heron Gate redevelopment as a long-term investment and that the current event was only the most recent in a series of community engagement initiatives undertaken by a benevolent and caring landlord. Gómez-Palacio would then make a more in-depth presentation with numerous slides outlining the visioning plan and various design concepts for the redevelopment. I collected all open house materials including the presentation slides as part of a broader archival research effort to trace ruling relations through their textual mediation. The public presentations and slides reveal much about the redesign of the community, the erasure of its current composition, and the replacement of its inhabitants through a more liveable, sustainable, redeveloped built environment.

Timbercreek officials have consistently attempted to reframe the mass eviction narrative through a focus on voluntary relocation. At the February 2019 open house, Rogers described the displacements as "impacts":

> One of the most important things we heard and one of the most challenging things about redevelopment are the impacts it has on the people that are directly impacted. We're trying to create a lot more housing on site, we're trying to create a lot more affordable, diverse, family-oriented housing, and unfortunately in the process some people have been ... impacted. We've done our best to try and help people find new homes and to help them with costs of moving, but at the end of the day nobody likes to leave their home.... Redeveloping property involves impacts like this.

Rogers' assertion that Timbercreek is trying to create a "more affordable, diverse, family-oriented" community is certainly debatable given that existing residents pay below-market rents in one of the most diverse neighbourhoods in the city, often in large, intergenerational families. The claim of creating an affordable community contradicts what Rogers said to Alta Vista residents in the September 2016 meeting, documented in Chapter 7, that the aim in the Vista Local redevelopment was to offer "premium" rents and smaller units that large Somali households, in particular, could not afford or fit in. While admitting that nobody likes to leave their home, Rogers suggested these types of impacts are inevitable when it comes to the business of redeveloping property.

During the March 2019 event, audience members pressed Timbercreek about which definition and metric of affordability it would apply to its new units. In a moment of honesty, Rogers awkwardly articulated Timbercreek's approach to affordable rental housing as a financialized real estate investment firm: "At the end of the day, if you're renting a place for less than market, you're doing stuff by definition that's not market, and what we do is market." When confronted with the fact that existing Heron Gate units rent at below-market prices, he said that in the end Timbercreek would be offering more affordable units than what had been demolished thus far. Opponents pointed out that the master plan would ultimately result in the loss of more affordable units than would be built; moreover, none of the new units would rent anywhere near below-market prices. At the time, Rogers' admission that Timbercreek was not in the business of, or interested in, offering below-market rental units forewarned of the final terms and metrics of affordability that would be outlined in the social framework and MOU as part of the Official Plan Amendment.

A Social Framework for Heron Gate

Timbercreek first introduced the idea of a social framework for the Heron Gate redevelopment at the February 2019 open house event. In the social framework, Timbercreek pledges to cause no further displacement (residents will have the opportunity to move within the community before their homes are demolished) and to produce new affordable units (although its affordability metrics have received widespread criticism). Print-out displays and slides, along with presentations delivered by Rogers and Gómez-Palacio, showcased these commitments (see Figure 9-1).

Redevelopment proponents present the social framework as a benevolent gesture that goes above and beyond what is legally or morally required of real estate developers. Councillor Cloutier has referred to the social framework as a social contract, and some groups have called it a type of community benefits agreement. All of these actors — Timbercreek, Cloutier, and the community groups — have taken credit for instigating what they consider a landmark agreement in Canadian rental housing redevelopment. Gómez-Palacio said at the February event that never in his career had he heard of developers committing to anything like this — "so let it sink in," he prompted the audience.

Timbercreek based its social framework on the Conference Board of Canada's Community Wellbeing Framework (see DIALOG 2022). The

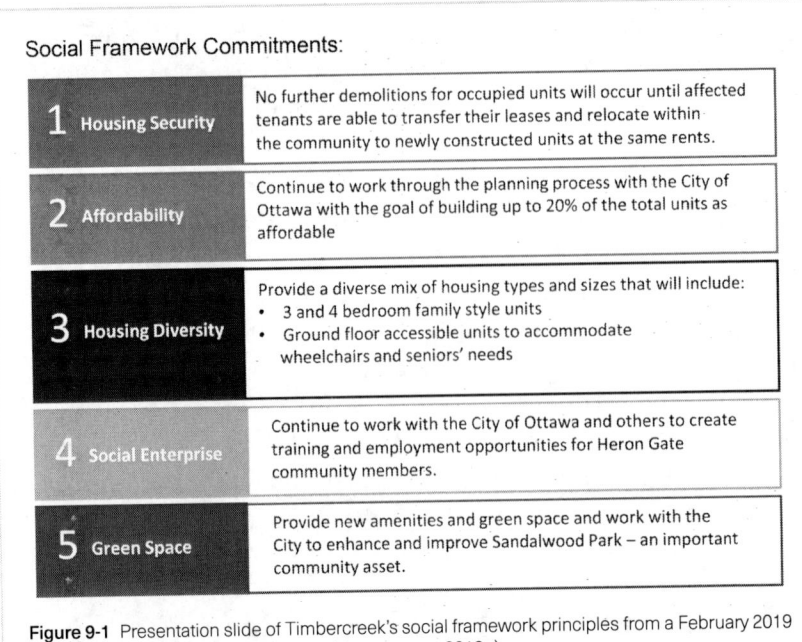

Social Framework Commitments:

1	**Housing Security**	No further demolitions for occupied units will occur until affected tenants are able to transfer their leases and relocate within the community to newly constructed units at the same rents.
2	**Affordability**	Continue to work through the planning process with the City of Ottawa with the goal of building up to 20% of the total units as affordable
3	**Housing Diversity**	Provide a diverse mix of housing types and sizes that will include: • 3 and 4 bedroom family style units • Ground floor accessible units to accommodate wheelchairs and seniors' needs
4	**Social Enterprise**	Continue to work with the City of Ottawa and others to create training and employment opportunities for Heron Gate community members.
5	**Green Space**	Provide new amenities and green space and work with the City to enhance and improve Sandalwood Park – an important community asset.

Figure 9-1 Presentation slide of Timbercreek's social framework principles from a February 2019 open house event (Timbercreek Asset Management 2019g).

Conference Board of Canada (2022a, 2022b), which describes itself as "the country's largest private economic analysis and forecasting unit," is an affiliated but independent operation of the Conference Board, Inc., which specializes in "business intelligence." The business-oriented outlook and intended outcomes of the framework, then, should come as no surprise. The framework originated with a research report coproduced by DIALOG and the Conference Board of Canada — *Community Wellbeing: A Framework for the Design Professions* (Markovich, D'Angelo, and Dinh 2018) — the stated purpose of which was to study the relationship between the built environment and the wellbeing of people (DIALOG 2022). According to the authors, the research was prompted by a growing interest in "design decisions that foster health, vitality, and wellbeing" (Markovich, D'Angelo, and Dinh 2018, ii).

A blueprint for revitalization and liveability, the Community Wellbeing Framework is premised upon intervening in and changing the built environment in order to facilitate healthier living. It is depicted on a concentric circle, with key indicators of wellbeing that include social, cultural, political, environmental, and economic dimensions. Timbercreek displayed this graphic in presentation slides during the

2019 open house events, emphasizing that Heron Gate would become the first community modelled after the Community Wellbeing Framework (Link2Build Ontario 2021).

Defining liveability and wellbeing

The foundational report for the Community Wellbeing Framework dedicates considerable space to defining what is meant by community wellbeing. Markovich, D'Angelo, and Dinh (2018, 11) discuss the synonymous relationship of wellbeing and liveability, adopting Lowe et al.'s (2013) definition of liveability as a space that is "safe, attractive, socially cohesive and inclusive, and environmentally sustainable; with affordable and diverse housing linked to employment, education, public open space, local shops, health and community services, and leisure and cultural opportunities; via convenient public transport, walking, and cycling infrastructure."

Further, the authors note the concept of liveability's concrete relation to community wellbeing and relevance to the design professions, as liveability "reflects the wellbeing of a community and comprises the many characteristics that make a location a place where people want to live now and in the future." This description derives from a draft report on liveability from the state of Victoria, Australia, which states, "Enhancing liveability is important not only from the point of view of the quality of life of existing citizens, but it also impacts on the competitiveness and future prosperity of the State" (Victoria Competition and Efficiency Commission 2008). Liveability's dual function is emphasized here, linking the wellbeing and quality of citizen life with state prosperity.

Community wellbeing is determined through particular understandings of liveability and quality of life, with the ultimate aim of enhancing growth, performance, and prosperity. Defined as "the combination of social, economic, environmental, cultural, and political conditions identified by individuals and their communities as essential for them to flourish and fulfill their potential" (Markovich, D'Angelo, and Dinh 2018, 13), this understanding of community wellbeing moves beyond considerations that centre quality of life and happiness, according to the report authors, in favour of a more multifaceted construct that better reflects the diversity inherent to the term. Interestingly, this definition is taken verbatim from Wiseman and Brasher's (2008) article "Community Wellbeing in an Unwell World," which urges considerations of community wellbeing to move beyond the prevailing metric of economic growth. A closer look at the *Community Wellbeing* report reveals, however, the central role that rentier capitalism and settler subjectivities play in considerations of liveability.

The business imperative: Wellbeing return on investment

The foundational report on community wellbeing approaches liveability in entrepreneurial terms, emphasizing economic growth, performance, and prosperity. Large portions of the report are dedicated to expanding these ideas and making the case for the potential economic benefits of applying the Community Wellbeing Framework to urban design.

Chapter 2 of the report is titled "Building a Business Case in Support of Designing for Wellbeing." It opens with a question: "Why is community wellbeing important from a business perspective?" (Markovich, D'Angelo, and Dinh 2018, 18). According to the report, community stakeholders and project investors ought to approach development and design through this particular lens in order to identify and realize the economic benefits that accompany community wellbeing. The authors examine project examples across five sectors — workplace, academic, hospital, retail, and residential — and argue that infusing wellbeing design into residential projects correlates with improved profits and property values. Designing for wellbeing will produce "more attractive, marketable neighbourhoods" and "translate into economic gains," as improved mental and physical health conditions increase "resident satisfaction" as well as "capital and rental values" (Markovich, D'Angelo, and Dinh 2018, 37). In other words, "healthier, more comfortable homes" are "more likely to have higher rental and capital value" (Markovich, D'Angelo, and Dinh 2018, 38). Further, enhanced tenant satisfaction will "support or improve a business's brand recognition and equity." Thus, the Community Wellbeing Framework approaches health as a method for creating and enhancing value while harnessing tenant satisfaction as a marketing tool.

Chapter 2 of the report also directs readers to an appendix for examples of calculated returns on investment where the design framework has been implemented to attain "the full potential of spaces" (Markovich, D'Angelo, and Dinh 2018, 17). This section of the report provides project case studies for each of the five sectors identified as capable of producing economic benefits. Six pages' worth of tables identify "where projects calculated a return on investment (ROI) for wellbeing" (Markovich, D'Angelo, and Dinh 2018, 136) across these five sectors. The tables include columns for "improved wellbeing outcomes"; "savings or increased revenue" or "return on investment," depending on the table. In the table for homes and residential communities, some of the improved wellbeing outcomes include higher property value, rental rates, tax revenue, and business revenue, along with decreased crime. For example, the table highlights increased rental revenues of 7 percent with improved landscaping aesthetics, and increased property values with enhanced scores for mobility and walkability (one of the defining features of the City of Ottawa's New Official Plan's focus on 15-minute neighbourhoods).

One final note on the return on investment tables: they overwhelmingly correlate amplified economic value with sustainable environmental design and "naturalized environments" across all sectors. With their focus on the physical environment and urban sustainability, frameworks seeking to enhance community wellbeing have been described as producing a form of "green gentrification" (Jelks, Jennings, and Rigolon 2021). In turn, research has highlighted gentrification's negative impacts on health and wellbeing (Iyanda and Lu 2021; Thurber and Krings 2021), providing further insight into how community wellbeing design seeks to produce liveable conditions for some but results in the removal of others. In the context of the Community Wellbeing Framework, liveability and wellbeing are discourses of improvement that are mobilized with a business imperative to both enhance and further commodify the social relations of property, shedding light on how community wellbeing will be applied to Heron Gate.

Governing through community wellbeing

As the *Community Wellbeing* report demonstrates, there is a business imperative behind designing for community wellbeing and liveability. However, this kind of design also involves shaping the conduct of populations so they adopt desirable lifestyles and avoid undesirable behaviours, essentially becoming more productive citizens.

The Community Wellbeing Framework's calculations of liveability and return on investment go beyond rental revenue and property value to also include worker productivity. For example, the report emphasizes the role of transportation in "increasing the quantity and quality of the lower-wage labour pool, which can reduce business costs and increase productivity and competitiveness. Improving affordable transport options also tends to expand the labour pool for industries that require numerous lower-wage employees, such as hospitality and light manufacturing" (Markovich, D'Angelo, and Dinh 2018, 142). This design for liveability not only accounts for economic value in terms of property enhancement but also includes a class dimension of creating and enabling flexible and cheap labour pools for the efficient functioning of capitalist enterprise. The report provides insight into evolving mechanisms of entrepreneurial neoliberal governance that centre and elevate property relations through ideologies of urban growth as well as class relations.

Community wellbeing design is not restricted to urban design and also encompasses governance, population management, and the shaping of conduct, where "the design of spaces and places can be used to promote better or healthier choices and constrain undesirable behaviours" (Markovich, D'Angelo, and Dinh 2018, ii). In this regard, the aim is to "[create] conditions that promote desirable lifestyle behaviours and foster healthy community living through design features" (Markovich, D'Angelo, and Dinh 2018, 19). The Community Wellbeing Framework seeks to mould the ideal neoliberal settler subject, akin to what Miller and Rose (2008, 92) refer to as "government through community," which involves building allegiance between citizens and communities through strategies of governance "in the service of projects of regulation, reform, or mobilization."

Governing through community, in the present case, means designing for community wellbeing to enable a "healthier and more productive life" (Markovich, D'Angelo, and Dinh 2018, 38), where improved health and wellbeing leads to enhanced productivity and more productive citizen subjects. Thus, the discourse of liveability is directed not only at the improvement of property but also at the shaping of settler subjectivities. The reproduction of urban space in settler societies works in tandem with shaping the behaviour of the people who will occupy that space. In the case of Heron Gate, nonpreferred tenants are evicted and replaced by a more preferred settler citizen subject. Through the remaking of Heron Gate, both liveable *and* disposable subjects are produced alongside domicide. Liveability is produced through community wellbeing design and new "liveable homes" for the affluent, while domicide is produced through the destruction of homes and once thriving social and cultural networks. Thus, community wellbeing is designed for the optimal, productive settler subject who can contribute to the improvement of property through homeownership or high rents.

"A Termination Plan": The Heron Gate Official Plan Amendment

To achieve its vision for community wellbeing at Heron Gate, Timbercreek sought to demolish 559 more homes. Ongoing demoviction was necessary to fulfill the master plan of intensifying the property with dozens of new apartment towers comprising thousands of new units. Thus, in 2019 Timbercreek submitted a proposal for its master plan to the City of Ottawa in the form of an Official Plan Amendment.

Urban development policy at the municipal level is largely directed through provincial legislation such as the Ontario *Planning Act* and *Provincial Policy Statement*, which mandates municipalities to create official plans (Government of Ontario 2020). Official plans provide a vision for the future growth of a city and guide the physical development and use of land over the long term. If landowners wish to develop their properties in a manner that conflicts with the provisions of an official plan, they can request an amendment. In the case of Heron Gate, Timbercreek wanted taller buildings and more units than the existing official plan allowed. In an application to the City of Ottawa, it asked for approval of a new site-specific policy for the area, which would, according to city planners, establish a vision, guiding principles, and various strategic policies pertaining to "land use, built-form, public realm, transportation, circulation, sustainability, housing, and community benefits" (City of Ottawa 2021f, 11).

Weeding through the technical language makes clear that an Official Plan Amendment is a land use tool that enables property developers to bypass regulations and limitations on development established by municipal policy and provincial law. Existing zoning allowed for a maximum of 9 storeys and 4,988 units. In its initial amendment application, Timbercreek sought as high as 40 storeys and 6,427 units; given intense opposition from Alta Vista homeowners to the north concerned about shadows cast by tall buildings, the landlord-developer eventually settled on a maximum 25 storeys, but was able to retain the original number of units by spreading out the height loss to other buildings throughout the property (see Figure 9-2) (City of Ottawa 2021e).

126 RESISTING EVICTION

Timbercreek submitted its initial Official Plan Amendment proposal, accompanied by a number of plans, designs, surveys, and assessments, to the City of Ottawa on April 24, 2019. More documents were submitted in February and December 2020. In all, around fifty documents associated with the application were produced and made available to the public. During this time, Timbercreek rebranded as Hazelview. Finally, on August 13, 2021, the planning department tabled a report and other relevant documentation, including an MOU with the landlord-developer outlining the terms of the social framework, for review by the planning committee. City council voted to approve the amendment on September 8, 2021.

The main forty-nine-page report submitted as part of the Official Plan Amendment provides the number of already demolished, existing, and to-be-demolished units. The report notes that 230 townhouses had been demolished thus far in the neighbourhood, making way for the construction of 348 units in Vista Local and rendering vacant another 6-hectare parcel of land (City of Ottawa 2021f). The current rental unit count was 1,864, which included 957 units in five apartment towers (ranging from 8 to 19 storeys), 307 townhouse units, 252 units in low-rise apartment buildings, plus the 348 Vista Local units. The report then acknowledges that the majority of units outside of Vista Local currently rent below the average market price and house numerous linguistically and racially diverse households, notably Arabic, Somali, and Nepali

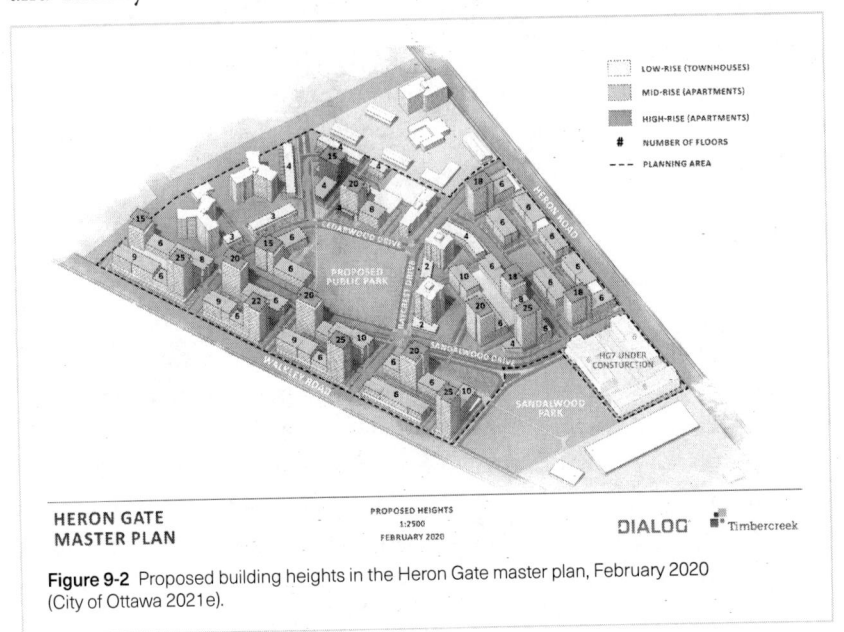

Figure 9-2 Proposed building heights in the Heron Gate master plan, February 2020 (City of Ottawa 2021e).

families. In terms of planned demolitions, the five existing apartment towers and their 957 units would remain, while the remaining 559 townhouse and low-rise units would be torn down. The MOU provides detail on the number of units to be built. In addition to the 1,305 existing units in the five towers and Vista Local, Hazelview would add another 5,122 new units, of which 1,439 were enabled through density bonusing (City of Ottawa 2021c). Density bonusing permits property developers to exceed the density limits in zoning bylaws in exchange for providing public amenities; in this case, it allowed Hazelview to raise the tower heights up to 25 storeys.

As a formal agreement about the social framework, the MOU was a unique point of contention and discussion. When Timbercreek introduced the principles of the social framework in 2019, it said it was doing so as a benevolent property owner that cared about the wellbeing of its residents. Meanwhile, Councillor Cloutier, who faced immense public pressure as the local elected representative of the ward where some eight hundred people had been evicted and dispersed, claimed that his office had initiated the social framework and demanded that the landlord-developer sign on to it. Community organizations such as ACORN applied consistent pressure tactics and also claimed credit for bringing about the conditions of the social framework. For instance, ACORN wanted a commitment to housing security, meaning no more displacement; instead, tenants should be able to transfer their leases and relocate within the community to newly constructed units at the same rents (ACORN Ottawa 2021). The organization also wanted to see up to 20 percent of total new-build units made affordable. In 2021 ACORN conducted a survey of 104 Heron Gate tenants on the social framework principles and found overwhelming support for affordable housing options and the right of already-evicted tenants to return. The survey results also showed that 86 percent of respondents had not attended a public consultation session (ACORN Ottawa 2021).

Timbercreek had to contend with the wrangling of multiple social and political actors when it came to the social framework. It insisted that the framework be one in principle only, not legally binding, arguing that the firm's public, verbal commitment would be sufficient, as a lot was riding on what it purported was the integrity of its brand. Intense negotiations between the landlord-developer, city planners, and Cloutier ensued, as gritty details on the number of and timeline for affordable units were ironed out. The landlord-developer, now under the name of Hazelview, signed the MOU on September 8, 2021. It restricts the City of Ottawa from attempting to secure any additional affordable units within Heron Gate as part of any future planning approvals or redevelopment. Facing passionate opposition after the 2016 and 2018 mass evictions, the landlord-developer has agreed to no demolition or renovation of occupied units until tenants are offered a newly constructed equivalent unit of the same type and rent (City of Ottawa 2021c). While this type of policy should be a legislated standard, it was only won through fierce grassroots tenant organizing and strong pressure levied by other groups.

The municipal planning vision for Heron Gate mirrors that of the developers. In the City of Ottawa's (2021d, 5) vision for the neighbourhood, "Heron Gate will be comprised of approximately 6,400 units and will be a complete, vibrant and sustainable 15-Minute Neighbourhood that enhances the quality of life for residents, while harmonizing with surrounding communities." The emphasis on harmonizing Heron Gate with Alta Vista, in particular, aligns with the landlord-developer's vision. Likewise, the first guiding principle is to "shape a vibrant community identity that enhances the quality of life for all" (City of Ottawa 2021d, 5). As articulated throughout this book, the efforts to revitalize Heron Gate have included the replacement of a particular community identity, that of a racialized, working-class neighbourhood, with something more affluent in line with Alta Vista demographics. The production of liveability — vibrancy, vitality, quality of life — requires gentrification, displacement, and the elimination of an ethnoracial enclave.

Attempted erasure at Heron Gate is further reflected in the City of Ottawa's amendment specifications for affordable housing. The municipal government "supports the development of a more inclusive and equitable community by maintaining a supply of affordable housing for low- to moderate-income households" (City of Ottawa 2021d, 5, 17). This phrasing implies that the government and developers are able to boost inclusivity and equity beyond what already exists on the site. The amendment and area-specific policy documents, as well as previous document submissions, contain original neighbourhood images as well as renderings and maps of the phases of redevelopment. In Figure 9-3, the first image shows what the area looks like before implementation of the master plan, the second image provides a sense of which parcels will be demolished, and the third image shows what the area will look like after.

Community Wellbeing: A Social Framework for Domicide 129

Figure 9-3 Images created in support of the Heron Gate master plan, showing the area before and after plan implementation
(City of Ottawa 2021d).

The devil in the details:
Affordable housing metrics and gentrification

Before going to city council for a vote, the Official Plan Amendment application required approval from the planning committee. A committee meeting was held online on August 26, 2021, and lasted all day. City staff and the landlord-developer made presentations. A number of public delegations, all of which were in opposition, also signed up to speak. At this event, as well as the city council meeting on September 8, 2021, I recorded audio, took notes, and gathered textual materials (agendas, planning documents, and presentation slides).

The City of Ottawa's planning committee heard from fifteen public delegations, which included local community groups, housing advocates, and current and former Heron Gate tenants. One tenant spoke powerfully about what they interpreted as the purposeful neglect of their home by the landlord:

> I'll tell you my experience has been quite horrible, and one of the key factors is in terms of maintenance of the property. The developer doesn't maintain the property. I've never lived in such a situation. I come from Africa, but I've never lived in such an unliveable situation anywhere else in the world. And to see people struggling in terms of getting facilities, in terms of getting maintenance service, in terms of having consultation with the developer is really deplorable. I don't want to believe these are the values that Canada is known for.... And I believe if we all come together and work together, we can find a lasting solution to the housing situation in Heron Gate.

According to the tenant, the landlord had rendered the neighbourhood unliveable through neglect and disrepair. They referred to the Official Plan Amendment as "a termination plan" to remove poor residents from the neighbourhood.

The major issues raised by the public delegations, along with progressive councillors opposed to the amendment, centred on how affordable housing would be measured and implemented. There was broad consensus among the range of opponents that the existing socioeconomic demographics in Heron Gate would be unable to afford the new affordable housing detailed in the MOU. The City of Ottawa's director of housing services, Saide Sayah, admitted in a presentation that the affordability measurements negotiated with Hazelview were not necessarily affordable for low-income households.

The final MOU stipulates that within the first five years of the approval of the Official Plan Amendment, Hazelview will designate 510 existing units as "secure affordable existing" units — described as units with rents "at or below average monthly City-wide rents by unit type" (City of Ottawa 2021c, 2) — for a period of twenty years. This metric of affordability defines affordable rents as current average market value as determined by Canada Mortgage and Housing Corporation figures: the average market rent for a two-bedroom apartment is pegged at $1,517 and a three-bedroom at $1,850. Since existing rents in Heron Gate fall below this determined average, "secure affordable existing" units will see rent increases. Through the MOU, the city and the landlord-developer have redefined affordable housing in a lower-income neighbourhood to be units with city-wide average market rents. Opponents at the planning committee and council meetings decried city planners for not negotiating Heron Gate–specific affordability measurements in line with current resident income levels, which fall far below the city average, as shown by the neighbourhood's own census tract.

In addition to the conversion of 510 existing units to "affordable" at-market rents, another 510 units from the redevelopment are to be designated as "secure affordable new" units for a period of ten years. This designation, and the resultant rent, is tied to 30 percent of average income based on city-wide average income percentiles. The MOU indicates that the average annual household income for a family living in a three-bedroom unit in Ottawa is $92,784; however, average household income in Heron Gate is $40,594. The actual average affordable rent in Heron Gate, based on the Hazelview formula of 30 percent of average annual income, is $1,015 per month (Mast and Hawley 2021). Meanwhile, existing tenants living on the property in three- and four-bedroom townhouses with front and backyards are currently paying around $1,500. Under the MOU affordability scheme and calculation based on an annual income of $92,784, new "affordable" three-bedroom apartments will rent for $2,320 per month, an increase of about 55 percent over current rates (Mast and Hawley 2021). Opponents further criticized the length of time that the units would remain "affordable." Mavis Finnamore of ACORN Ottawa — who was displaced in the 2015–2016 demoviction — put it this way to the committee: "The poor are always going to be with us. Why are we pretending they are going to evaporate in ten to twenty years?" For Finnamore, the Official Plan Amendment finalized the "blowing apart" of a community, including support systems for new Canadians and the removal of housing security.

Affordability in a renewed and revitalized Heron Gate will be unaffordable for many current residents, who rely on rents below the market average. Moreover, as more and more parcels in Heron Gate are redeveloped and above-market renters move into the new buildings, rent gaps will impact the immediate neighbourhood and wider community. Both Hazelview and surrounding landlords will seek to close these gaps in rent through rent hikes, AGIs, and evictions. More racialized and lower-income people will be displaced by processes of gentrification that accompany the larger redevelopment. While existing tenants will be able to relocate within the neighbourhood at the same rent, according to the MOU, the overall percentage of units that are marked as "affordable" is relatively small. Moreover, these units will only be priced as such for a decade or two, a generation or less in a neighbourhood that is home to many multigenerational families. As the Herongate Tenant Coalition and others have emphasized, Hazelview's redevelopment will destroy much more affordable housing than it will ever create.

"A home run": City planners go to bat for Hazelview

For a handful of more progressive-oriented inner-urban councillors, the affordability clauses made little sense. Councillor Shawn Menard, who opposed the Official Plan Amendment, hit Hazelview representatives with some tough questions at the planning committee meeting. He wanted to know if the firm intended to tear down the property when it made the purchase in 2012, and if it did its homework and knew that approximately 70 percent of the neighbourhood was made up of lower-income, racialized immigrant families. When Menard failed to receive a direct answer, he made the following statement to Hazelview's representatives, Michael Williams and Colleen Krempulec:

> The selection of Heron Gate for redevelopment was not fortuitous. Generally, racialized and immigrant communities like Heron Gate are disproportionately likely to be selected by large-scale redevelopment projects and thus subjected to mass evictions. Further results suggest that the dissolution of the Heron Gate community, and the attendant dislocation of

its members has exacted a pronounced social and economic toll and compounded the racial discrimination already experienced by the former Heron Gate residents, most of whom are visible minorities.... You're an example of a black eye for this country, and I'd like you to respond to the people that have been calling you out and rightfully for what is racial discrimination so that you can profit off of it.

Krempulec, senior vice president of brand marketing and corporate social responsibility at Hazelview, responded that the predominant reason that the firm was engaging in redevelopment was to add affordable housing supply to the city: "There is no racial discrimination here whatsoever. The merits of the application are based on the need to bring much-needed housing, rental housing, across various affordability levels to not only the city of Ottawa but to the Heron Gate community. And that's what we're here to discuss today."

Menard replied:

> If that's what you're here to discuss, then why is the MOU so weak? Why does it actually say it's ... affordable when the definition of affordable isn't actually affordable for the people that live in this area? Or, why is the length of time ten to twenty years and fifteen to twenty years when we have developments at Booth [Street] that are twenty-five years, developments in Toronto that are ninety-nine years? If that's your intent to come in and say that you really just want to build new housing and be supportive of this community, then why is the MOU so weak and the very tenants that you're saying you want to protect are upset by the deal and agreement that you put in place here? So don't pretend that you're doing this out of the goodness of your heart.

Krempulec's claim that Hazelview was redeveloping Heron Gate to add much-needed affordable housing to the city's rental stock was bold, given Greg Rogers' public meeting statements in 2016 and 2019 that the firm was not in the business of providing affordable housing and uninterested in renting to the larger families living in the townhomes.

The landlord's public position had changed in lockstep with municipal planners and city officials who supported the Official Plan

Amendment, where the selling point hinged on adding housing to the market. Any further delay to approving the amendment, the argument went, would delay the addition of vital rental housing in the tight Ottawa market. However, letting a developer build housing does not necessarily mean enhanced affordability will follow, as developers often build for the high end of the market (as seen in Vista Local), not for the greatest need. Increasing supply at the top does not reduce demand at the bottom (Stein 2019). What's more, a greater number of luxury units is likely to drive up neighbouring rents and property values; in the case of Heron Gate, redevelopment will lead to wider community gentrification as neighbours in the surrounding areas are priced out through elevated rents. While Menard and two other committee members saw through Hazelview's exercise in brand marketing, the planning committee voted six to three in favour of the city planners' recommendations.

At the city council meeting, Hazelview's representatives had support from the City of Ottawa's planning department. In particular, Lee Ann Snedden, the director of planning services who officially submitted the Official Plan Amendment recommendation report, stepped up to bat for the developers. At times it was difficult to know if Snedden was speaking as a public servant or a Hazelview representative. "To expect that a private developer is going to provide affordable housing in perpetuity is unrealistic. It's completely outside of the *Planning Act*," she said at one point, responding to criticism of the ten-year timeframe for new affordable units. When councillors pressed Snedden on the affordability metrics, she replied with incredulity, "They don't have to provide us *anything*. We've never … we've got over *1,000 units* with this agreement. This is a landmark. This is a *home run* with respect to affordable housing and the number of units that we're providing."

Some councillors remained unconvinced and pushed for a better deal. In response, Snedden remarked that Hazelview "can only take so much in terms of what they are able to financially able to afford to provide." Anything beyond the MOU, she said, "would be going past what they can afford to build here." Snedden's posturing during the exchange with city councillors reflects the symbiotic relationship between municipal planners and property developers. City staff advocacy for the Heron Gate developer summoned disbelief from skeptical city councillors such as Jeff Leiper:

Sorry, I'm just struck by the previous comment by Ms. Snedden in terms of, you know, the affordability of this deal for the developer. We're not talking about what is affordable to the developer. This is what the developer feels that they can accomplish while still making the profit that they intend to make in this development. We're talking about affordability of housing to people who actually can't afford to pay more than they are paying for their housing today versus the affordability of whether or not profit can be maintained at the desired level on the part of investors....

There is a housing shortage in most major North American municipalities, and in the wake of that housing shortage, there is the opportunity for profit, and large companies like Hazelview are swooping in and they're buying what they tell their investors are underperforming assets and they are squeezing greater profit out of them by redeveloping them. And the message that we have the opportunity to send to those developers today in this debate is that Ottawa City Council, Ottawa city staff are going to try to do better....

We have the opportunity to say our cities are not going to be bought up by large developers backed by billions of dollars of investment to turn housing that is affordable for people today into a commodity from which the greatest possible profit can be squeezed. And when we talk about "is this deal affordable, if this deal is affordable to the developer," that leaves me queasy. We're not talking about the same affordability. We're talking about the affordability of housing, which is a human right to the residents of Ottawa versus the profits that can be made by speculative large multinational firms.

While Snedden framed the deal with Hazelview as constrained by the landlord-developer's need to realize adequate returns on investment, progressive inner-urban city councillors were not satisfied. They felt strongly that the city had a strong position from which to negotiate better terms in the MOU, given that the New Official Plan for Ottawa was to be approved in the coming weeks and would provide stronger mechanisms compelling developers to include affordable units. At one point, Snedden retorted, "We don't have the staff capacity to continue these negotiations and discussions." She observed the city had twenty-three wards and planning staff were busy with other files and applications, the latter of which had risen by 20 percent. "This has been the strongest agreement that we've been able to acquire," she insisted. From the planning department's perspective, going back and trying to negotiate a better deal was out of the question. Despite the best efforts of opposing city councillors who wanted a stronger MOU, city council passed the amendment with a vote of eighteen to six.

Despite strong public opposition to the social framework and MOU, a majority of Ottawa city councillors were convinced to support the Official Plan Amendment. It appeared that planning staff had effectively pleaded the case for the redevelopment. A city councillor staff person that I interviewed explained from their perspective the role of urban planning in the approval process:

> The planning department is basically in the business of facilitating investment for developers. That's kind of what they're doing is they're like grooming plots of land and planning it and facilitating investment there. And the city, you know, is reliant on that, that industry continuing to do well. And they're reliant on continuing to facilitate that capital strategy because that's where the bulk of their revenue comes from for the city, in property taxes. (Anonymous interview, June 18, 2021)

The relationship between developers and municipal governance is mutually beneficial, as city staff work within an urban planning bureaucracy designed to prep property for investment and improvement.

The municipal-developer nexus

The municipal-developer nexus — the symbiotic relationship between city officials and property developers — is built around the desire to obtain and improve property and boost revenue flows. Property relations are central to capital flows that enrich landlords and their investors and provide cash-strapped municipal governments with a main source of revenue. This is part of what Stein (2019) classifies as the "real estate state," a political formation and expression of governance that makes real estate a top political priority, a primary commodity, and a dominant revenue stream for cities. In the real estate state, urban governance is increasingly preoccupied with commodifying land and upholding property rights and relations as sacred. Urban planners "are tasked with anchoring municipal governments to the interests of developers and landlords" (Stein 2019, 114–15), which ultimately results in "maintaining the spatial dimension of racial inequalities" in urban property relations (Stein 2019, 27). Developers enjoy a privileged position within the planning bureaucracy, gaining access to land for the purposes of development and redevelopment.

While financially beneficial for both sets of parties, urban development is also associated with improving and beautifying settler society. The reproduction of urban space, central to settler colonial urbanism, relies to a significant extent on the municipal-developer nexus. The improvement of land appeals to broader settler society, facilitated by discourses of liveability that lead to normative processes of gentrification. While numerous factors influence the praxis of developers and apartment investors in the built environment, including social forces pressuring for concessions, mechanisms of urban governance facilitate developer access to property and its improvement.

There is widespread general mistrust of the relationship between municipal officials (councillors, planners, etc.) and property developers. Municipal officials openly admit this during interviews, and city reports from public engagement efforts capture a broad public perception that developers enjoy a cushy seat at municipal planning and decision-making tables while the general public is largely shut out. Part of my work with the Herongate Tenant Coalition has been to investigate and expose the municipal-developer nexus and societal powerbrokers that feel they can dismantle communities and reap enormous profits with ease and impunity. To this end, I filed a series of FOI requests to enhance my understanding of the City of Ottawa's planning department in relation to property development.

One particular program I investigated involved matching developers with an ambassador in the planning department. Before her appointment as the director of planning services in 2017, Snedden chaired the City of Ottawa's Development Industry Steering Committee. At the time of her promotion, developer media said she had "a good reputation with the building and development industry" (CNGRP 2017). One of Snedden's first initiatives in her new position was to establish a pilot ambassador program, where private developers are assigned a city staff ambassador to facilitate and streamline the property development application process.

These staffers, referred to as "client relationship leaders," have the authority to "override decisions by project managers or the city staff member previously tasked with overseeing a particular planning file" (Pearson 2015). Internal planning documents describe the program as a "business process improvement initiative that will provide one point of contact for the most active development firms" (City of Ottawa 2021-549). Program goals include improving internal municipal planning communications and external communications with developers, as well as removing inefficiencies and accelerating timelines for application processing. At least twenty-three developers had been assigned ambassadors at the time.

City councillor and planning committee chair Jan Harder publicly backed the program. "The communities have ambassadors and they're called councillors," she retorted in the face of criticism that the city favoured developers over citizens and community groups. "The No. 1 goal here is customer service and moving applications through the system in a timely fashion" (Pearson 2015). Given her extensive history in municipal politics, planning, and property development circles, it is not surprising that she went out of her way to defend the ambassador program. When the Herongate Tenant Coalition's small team of researchers convened in 2018 around the mass evictions, Harder was one city official we wanted to learn more about. Our research drew connections between Harder and Jack Stirling, head of a development consultancy firm called The Stirling Group. Stirling had worked for large property development firms and then chaired the Nepean planning commission in the late 1990s (before the municipality of Nepean became part of Ottawa), when Harder was a Nepean councillor regularly seeking development approvals (Rockwell 2018b, 2019). We were first alerted to the possibility that Stirling was involved in the mass demoviction and redevelopment plan for Heron Gate when Mayor Jim Watson unexpectedly dropped his name in a meeting with community members in 2018. It came to light that The Stirling Group had submitted a site plan control application in relation to the first demoviction phase and Vista Local redevelopment. Then, at the February 2019 open house event, I picked up a printed copy of the presentation slides and noticed that, in addition to branding from the usual partners — Timbercreek, the City of Ottawa, and DIALOG — The Stirling Group's logo now appeared. The company had been tasked with submitting the Official Plan Amendment application to the city on behalf of Timbercreek.

While none of this is shocking or revelatory, our early coverage in *The Leveller* newspaper in 2018 and 2019 later bore fruit. First was Neal Rockwell's (2018b) article "Disconnected Realities: Country Club Politics Put Vulnerable Herongate Residents in Peril." Well-known former public officials had formed development consultancy firms and used their "insider knowledge and important connections" to assist developers with their applications. The article had a particular focus on Stirling and his position on the city's Planning Advisory Council. Then came a follow-up piece from Rockwell (2019) called "Herongate Residents Contend with Below-Grade Conditions and a Rent Increase, while Councillors and Developers Mingle." The subtitle read: "Broken heating, broken pipes and a rent increase in Herongate, while councillor Jan Harder and development consultant Jack Stirling keep it in the family." Within, Rockwell revealed that Stirling's daughter, Alison Stirling, was working as an aide for Councillor Harder while also employed by The Stirling Group. This particular article struck a chord with a local reader, who filed a complaint with the city's integrity commissioner.

Integrity Commissioner Robert Marleau conducted an inquiry, and on June 17, 2021, issued a report to city council. He found that Harder, as city councillor and planning committee chair, had violated the code of conduct for councillors and "tainted" the City's planning and development process through her business relationship with the Stirling family. In addition to hiring Alison Stirling, Harder had several times contracted The Stirling Group to advise her in her role as chair of the planning committee. Meanwhile, The Stirling Group represented private clients at the committee. Between contracts, Harder received unpaid services from the company. Marleau recommended Harder's removal from the planning committee as well as the Planning Advisory Committee and the board of directors of the Ottawa Community Lands Development Corporation, which oversees the sale of municipal lands to private parties. Harder resigned from the planning committee before a council vote on the matter, and city council voted down other proposed sanctions against her at the behest of Mayor Watson. The "Watson club," a relatively solid group of majority non-inner-urban city councillors that tended to vote for the mayor's recommendations, was in action, just as it had been with the Official Plan Amendment vote (Chianello and Porter 2021a, 2021b).

For his part, Stirling was sanctioned for violating the Lobbyist Registry By-law and the lobbyist code of conduct, which stipulates that a lobbyist with active files cannot offer gifts or benefits to council members or their staff. Stirling had provided free services to a councillor between November 2019 and February 2020 while he had three active lobbying files with the city. He was forced to sign a compliance agreement and was banned from lobbying for thirty days (City of Ottawa 2021g). While both Stirling and Harder received slaps on the wrists for their arguably serious transgressions, these penalties had symbolic importance and were the result of the Herongate Tenant Coalition's exposure of the linkages between actors in ruling institutions.

Part of the political activist ethnographic work in this research project is direct engagement in the social relations of resistance, which includes upsetting, disrupting, and changing ruling relations. This research has illuminated the inner workings of ruling institutions comprised of an entanglement of actors who seek to carve up the city for profits, as the earliest surveyors in Gloucester Township did. The Algonquin Anishinaabe land now known as Heron Gate was once surveyed by settlers and incorporated into Gloucester Township as private property. The 2021 approval of the Heron Gate redevelopment serves to reproduce settler colonial urbanism through a community wellbeing redesign that erases an ethnoracial enclave and reconfigures private property into whitespace for ideal settler subjects.

The Herongate Tenant Coalition's approach has been to largely disengage from the "consultation" and public relations exercises and negotiations over the ongoing destruction of the neighbourhood. The coalition interprets the Official Plan Amendment and MOU as a framework for destruction, used to justify the demolition of hundreds more homes and elimination of affordable housing in the community. While the landlord-developer and city officials have painted themselves as taking the moral high ground through the social framework, the catalyst for this agreement, in my view, was the grassroots anti-eviction efforts of tenants and their supporters in 2018. The organizing work of the Herongate Tenant Coalition, the drive to stop the evictions, the gruelling social media conflict and subsequent legal battles, and the human rights case concerning those displaced in 2018 cast a spotlight on the demolition-driven evictions and compelled the landlord-developer to modify its approach to redevelopment. While the majority of political agents and community actors involved in the Official Plan Amendment process pursued a gentler form of domicide at Heron Gate, the Herongate Tenant Coalition has remained steadfast in its politics of refusal.

CHAPTER 10

Human Rights and Racial Discrimination in Housing

Depending on whom you ask, Heron Gate is a revitalization success story and beacon for affordable housing provision, or a resounding loss of 789 affordable (below-market rent) homes, hundreds of residents, and an ethnoracial enclave. The approval of the Official Plan Amendment seems to have sealed the fate of the neighbourhood, greenlighting mass demolition and redevelopment and accelerating the gentrification of the wider community. The Herongate Tenant Coalition has refused to throw in the towel, however, and is still organizing against the community's dissolution. While the majority of coalition members who were active in 2018 are no longer so — as they were evicted and dispersed from the neighbourhood — a small core group continues to work in the community. One area of continued struggle is in the legal arena. This concluding chapter examines the human rights case filed by evicted tenants within the context of radical tenant organizing, while revisiting some of the main themes explored in this book.

A Precedent for Housing Rights

Timbercreek's efforts to engineer a population transfer and replace racialized tenants with "premium" white renters constitute discrimination on racial grounds and violate the right to housing in international law, argue Heron Gate evictees in a case currently before the Ontario Human Rights Tribunal (Yussuf et al. v. Timbercreek 2019). In August and September 2019, thirty-seven tenants evicted in 2018 filed human rights claims against the City of Ottawa and Timbercreek Asset

Management Inc. (and its affiliated companies Mustang Equities Inc., TC Core GP, and TC Core LP).

At the centre of the case is the targeted destruction of an ethnoracial community (Mensah and Tucker-Simmons 2021). The families demovicted in 2018 were dispersed throughout the city, many relocating to the suburban margins where rents are more affordable. Displacement has disrupted their social networks and cultural supports. In a novel and potentially precedent-setting move, a number of human rights applicants, the vast majority of Somali origin, are seeking a right to return to units of similar size and rent in the neighbourhood, as well as $50,000 each in damages. They want the tribunal "to determine whether a landlord has the right to displace a large group of residents of a low-income, family-oriented, racialized and immigrant community in order to create a predominantly affluent, adult-oriented, white and non-immigrant community in its stead" (Yussuf et al. v. Timbercreek 2019, 4).

As discussed in Chapter 5, large numbers of asylum seekers and immigrants sought refuge in Heron Gate beginning in the early 1990s; in particular, hundreds of Somali families fleeing civil war made the community their home. More recent migrants may have initially been attracted by the large homes and affordable rents, but newcomers from African, Arab, and South Asian countries also chose to live there "in order to integrate [into] the increasingly tight knit migrant and ethnic community that was taking root. They sought to live in close proximity to those with whom they shared certain personal characteristics such as ethnicity, culture, language, place of origin, and religion" (Yussuf et al. v. Timbercreek 2019, 9). By the time of the 2016 Canadian census, more than half of Heron Gate residents were immigrants and 70 percent were visible minorities, the median employment income was under $20,000 per year, 93 percent of residents were renters, and 28 percent of households received some form of public assistance (Yussuf et al. v. Timbercreek 2019).

Heron Gate evictees showed a sociological and spatial understanding of the neighbourhood in their application:

> It is not merely the high concentration of people from specific ethnic groups, but the resulting emergence of cultural institutions, social relationships, and ethnic businesses that cater to the specific needs of the community or communities that

distinguishes an ethnic enclave from a community that merely has a high concentration of residents from one or a small subset of ethnicities. (Yussuf et al. v. Timbercreek 2019, 10)

Ethnoracial enclaves offer lower-income minority and recent immigrant households opportunities to access informal services and supports in their first language as well as culturally relevant businesses. These opportunities are significant, the application argues, "given the acute dearth of economic opportunities available to ethno-racial minorities in light of growing xenophobia and Islamophobia in Canada" (Yussuf et al. v. Timbercreek 2019, 12). Enclaves further provide "a level of defense or 'cushion' against racial discrimination emanating from mainstream society. That is, they act as a bulwark against discrimination in Canadian society, affording their members greater security both literally and in a socio-economic and cultural sense" (Yussuf et al. v. Timbercreek 2019, 12). The hostility directed toward racialized groups and immigrant communities in settler society is driven by a xenophobic culture informed by political and social logics of elimination. Settler society's insatiable quest to obtain and improve property informs the eliminatory logics targeting ethnoracial enclaves like Heron Gate.

The human rights application also targets the business model of the landlord-developer as a financialized real estate investment firm. It situates the Heron Gate redevelopment within the larger trends of the financialization of rental housing and hypergentrification targeting lower-income, racialized areas:

> Timbercreek's development plan for Herongate is consistent with a broader model of real estate development in Ontario that disproportionately affects people of colour, immigrants, people receiving public assistance, and families. This development model involves identifying real estate that is 'undervalued', displacing the existing 'low quality' occupants, renovating or building higher-end rental housing or condos, and then marketing them to 'higher quality' tenants. The replacement tenants are disproportionately white and non-immigrant, with smaller or no families, and not receiving public assistance. (Yussuf et al. v. Timbercreek 2019, 5)

Value and quality are two underlying principles driving gentrification in the financialization of rental housing. Undervalued properties inhabited by tenants determined to be of low quality (or undesirable) are targeted for elimination through the injection of value and replacement by high-quality (or desirable) tenants. This eliminatory mechanism of settler colonial urbanization works to replace undesirable (typically nonwhite) populations with more desirable (typically white) populations, and is facilitated by processes of settler colonial urbanism and the reproduction of space.

For its part, Timbercreek/Hazelview denies the allegations and asserts that the case has no public interest standing and no reasonable prospect for success. According to the landlord-developer's response, filed with the Ontario Human Rights Tribunal in September 2019, the neighbourhood was not targeted because it contained lower-income, racialized families: instead, the "decision to redevelop the property is in keeping with the cycle of evolution and rejuvenation that is common in communities across the Province and across Canada" (Timbercreek Asset Management 2019h, 30). But the evictees assert that the redevelopment plan for Heron Gate violates the right to housing found in international human rights law, which has been ratified by Canada and is binding on all levels of government. As a result, the applicants argue that "displacing existing residents, particularly those who are low-income and members of groups experiencing discrimination, for the purposes of redevelopment, is only permitted if existing residents are fully consulted about development plans, and if those plans avoid permanent displacement of residents" (Yussuf et al. v. Timbercreek 2019, 5–6). In contrast, Timbercreek insists the case is not one of housing discrimination because "the right to housing has not been enshrined in domestic law" (Yussuf et al. v. Timbercreek 2019, 35).

The strength and merit of the case brought by evicted Heron Gate residents could be the push needed for housing to be legally recognized as a human right in Canada. As one of the most significant housing cases that the country has seen in twenty years, according to lawyer Daniel Tucker-Simmons, it has "the potential to restrict the long-standing and discriminatory practice in the real estate development industry of targeting racialized communities for displacement and gentrification" (Herongate Tenant Coalition 2021b). Yavar Hameed, another lawyer representing tenants displaced from Heron Gate, notes,

"The deck is always stacked in favour of property owners. Canadian law typically privileges the interests of capital above people. However, even property developers must abide by human rights laws. That is the strength of the Heron Gate case, which identifies the displacement of racialized tenants as a form of discrimination" (Herongate Tenant Coalition 2021b).

The Strategic Value of Legal Engagement

Engaging with a justice system that upholds a settler colonial legal framework that was built upon the sanctity of property rights can be fraught, to say the least. The conflict over Heron Gate provides insight into how tenant justice and urban social movements may benefit from a tactical and strategic engagement with the settler colonial (in)justice system. The various legal cases involving Heron Gate demonstrate that the legal terrain outside of landlord-tenant tribunals can be a viable option in the struggle against ruling institutions such as landlords, developers, and city officials. Housing activists widely recognize that landlord-tenant tribunals favour landlords and rarely deliver tenant justice. This is why elevating legal struggle beyond such tribunals can be fruitful if certain conditions are in place (e.g., competent and sympathetic lawyers, access to funding, and the time, patience, and energy to engage in the process).

The legal cases initiated by the coalition and Heron Gate tenants provide further insight into ruling institutions and social relations surrounding property, real estate, and municipal governance. The Herongate Tenant Coalition's tactical legal engagement is an exercise in knowledge production, a research strategy that enables tenant movements to better understand their adversaries and how they think about tenant organizing and resistance. We have learned that landlord-developers will go to great lengths to protect their brand; in fact, brand integrity is more important to them than any associated legal costs. Landlord-developers are susceptible to public shaming, and using social media against them helps level the playing field. Tenant efforts in this regard, in my view, pushed Timbercreek to reemerge as Hazelview.

Finally, if the human rights case is successful, Heron Gate tenants will have set a national precedent, possibly with international implications, that could provide protections from predatory real estate investment targeting buildings and neighbourhoods for gentrification. At the provincial level, a favourable ruling from the Ontario Human

Rights Tribunal could serve as the impetus for legislative changes that would protect tenants from demoviction and provide a right of return.

Housing Futures: Resisting Domicide in the Liveable City

Domicide is an incisive framework through which to analyze the concept of liveability and its application to Heron Gate. As a reproductive and repressive sociospatial process, domicide manifests under a discursive banner of liveability, where liveability for some depends upon domicide for others. It is repressive in rendering one group of people disposable and destroying their homes or community, and at the same time reproductive in replacing those people with a more privileged group under the auspices of improving quality of life and wellbeing. Attempts by Timbercreek/Hazelview and city officials and planners to make Heron Gate liveable for some involve unmaking the homes of existing residents whose dwellings and lives have been rendered unliveable. The already precarious lives of lower-income, racialized, and immigrant populations are made disposable so that other more desirable populations — typically representative of the white settler majority — can live and flourish. At Heron Gate, an entrepreneurial framework of community wellbeing has facilitated the improvement of property for settler use.

The remaking of Heron Gate has been made possible in part through ideas of liveability from Ottawa's New Official Plan and the Conference Board of Canada's Community Wellbeing Framework. The New Official Plan mobilizes liveability alongside intensification with the aim of "injecting new life" into targeted urban areas (City of Ottawa 2019b). The Community Wellbeing Framework mobilizes liveability alongside a business imperative to improve property relations, reproduce urban space, and shape productive settler subjectivities. The invocation of liveability within these particular frameworks of governance has both racial and spatial implications. The application of the Community Wellbeing Framework to Heron Gate reduces the racialized working-class neighbourhood to a space and place devoid of liveability. Designing for liveability involves processes of home unmaking, while designing for community wellbeing serves the purpose of modifying undesirable behaviours and generating productive life. These processes provide insight on how settler colonial urbanism works to reproduce urban spaces and ideal subjects in the development of liveable cities.

The Community Wellbeing Framework actually leaves the landlord-developer between a rock and a hard place, according to the Herongate Tenant Coalition. First, although the framework was released in July 2018, Timbercreek did not apply it at the time to the existing neighbourhood parcel subjected to mass eviction and demolition. Nor did Timbercreek apply it to the remainder of the neighbourhood (Mast and Hawley 2021). Using the Community Wellbeing Framework's metrics to evaluate Heron Gate in terms of social, cultural, and affordability components would likely result in a high grade, revealing a highly liveable neighbourhood for its residents. This would contradict purported justifications for demolition and redevelopment by undermining landlord-developer claims that the existing neighbourhood is "no longer viable" and exposing what the coalition has emphasized from the beginning: "viability" is synonymous with profitability. On the other hand, giving Heron Gate a failing grade despite high levels of sociocultural supports and affordability would reveal inherent (and potentially racialized) flaws in the framework and lay bare the framework's raison d'être of enhancing profitability and productivity through community design and redevelopment.

This book has examined two competing visions for the Heron Gate neighbourhood in Ottawa. The Herongate Tenant Coalition is struggling to maintain the working-class integrity of the neighbourhood for a lower-income, family-oriented, largely racialized and immigrant community. On the other side, Timbercreek/Hazelview's corporate capture of the community has initiated a process of accelerated financialized gentrification. The landlord-developer has employed demoviction as a blunt instrument of mass displacement to create a zone of affluent consumption for a higher-income, adult-oriented, white, and nonimmigrant community. Its efforts to demobilize the Herongate Tenant Coalition further demonstrate the implications of finance-driven gentrification and the attempted suppression of tenant organizing. Contested spaces such as Heron Gate represent strategic battlegrounds under settler colonial urbanism and the financialization of rental housing. The ongoing battle between Timbercreek/Hazelview and the Herongate Tenant Coalition is a localized but crucial flashpoint in contestations surrounding affordable housing and urban space in Canada.

Acknowledgements

Completing this book would not have been possible without the loving support of my family, friends, and colleagues. It would also not be possible without ongoing processes of settler colonization and Indigenous dispossession, to which my family has contributed through migration, settlement, and unhindered movement through the Mi'kmaw homeland (Mi'kma'ki) and my personal migration to the Algonquin Anishinaabe homeland, where I have been able to obtain an education and raise a family as an uninvited guest.

This book is a spin-off of doctoral research I completed at Carleton University (2018–22) in the Department of Sociology and Anthropology. This work was supported by my two PhD supervisors, Aaron Doyle and Jeff Monaghan, my committee members, Jackie Kennelly and Jen Ridgley, and my internal and external examiners, David Hugill and Ted Rutland. My pre-PhD and departmental pals Alexis Shotwell and Chris Dixon also deserve a special shout-out and warm thanks, as do Alex McClelland and the Tracking (In)Justice crew. Outside of Carleton, I would like to thank Kevin Walby and Martine August for their inspiration and support.

The amazing and dedicated team at Fernwood helped transform my research project into a publishable manuscript. In particular, Fazeela Jiwa and Erin Seatter were instrumental in bringing the book into its current form.

Although my larger networks of friends have dwindled over the last decade as I have refocused my energies on my immediate family, I am deeply indebted to a large group of people that have contributed to my personal and intellectual growth over the past twenty years. This list is far too long to elaborate on, but if you are reading this, you know who you are. I could not have undertaken this journey without the comradery and support of Jen, Guillaume, Ajay, Fuzz, Gassner, Thomas, Jakob, Birch, and Steve. Finally, I am grateful for the local punk and broader activist community in Ottawa and the opportunity to participate in such rad scenes with great people who are a source of unrelenting inspiration and fun.

I could not have undertaken or completed this project without the Herongate Tenant Coalition and the great folks that I have met through it. There are many to whom I owe a debt of gratitude, but in particular I would like to thank Josh, Nima, Tammy, Mumina, Ikram, Lily, Neal, Jocelyn, Nick, Daniel, and Yavar.

Thank you to my loving and supportive family, as well as my extended family and in-laws. In particular my sister, mother, and father provided significant support and guidance. Last but not least, and certainly the most, words cannot express my love and gratitude for my partner and children — Rachel, Èvie, and Noélie — who offer me an endless source of inspiration, love, and joy, for which I am forever thankful.

References

Aalbers, Manuel. 2016. *The Financialization of Housing: A Political Economy Approach.* London and New York: Routledge.

___. 2019. "Financial Geography II: Financial Geographies of Housing and Real Estate." *Progress in Human Geography* 43, 2.

ACORN Ottawa. 2021. "The Alternative Vision for Herongate: The Fight for the Right to Housing and No Displacement." Herongate ACORN Tenant Union.

Akesson, Bree, and Andrew R. Basso. 2022. *From Bureaucracy to Bullets: Extreme Domicide and the Right to Home.* New Brunswick, NJ: Rutgers University Press.

Alexander, Donald. 1970. "Liveable Cities." *Current History* 59, 348.

Amin, Ash. 2007. "Re-Thinking the Urban Social." *City* 11, 1.

Anderson, Elijah. 2015. "The White Space." *Sociology of Race and Ethnicity* 1, 1. https://doi.org/10.1177/2332649214561306.

Auger, Deborah A. 1979. "The Politics of Revitalization in Gentrifying Neighborhoods: The Case of Boston's South End." *Journal of the American Planning Association* 45, 4.

August, Martine. 2014a. "Challenging the Rhetoric of Stigmatization: The Benefits of Concentrated Poverty in Toronto's Regent Park." *Environment and Planning A: Economy and Space* 46, 6. https://doi.org/10.1068/a45635.

___. 2014b. "Negotiating Social Mix in Toronto's First Public Housing Redevelopment: Power, Space and Social Control in Don Mount Court." *International Journal of Urban and Regional Research* 38, 4. https://doi.org/10.1111/1468-2427.12127.

___. 2016. "'It's All About Power and You Have None': The Marginalization of Tenant Resistance to Mixed-Income Social Housing Redevelopment in Toronto, Canada." *Cities* 57. https://doi.org/10.1016/j.cities.2015.12.004.

___. 2020. "The Financialization of Canadian Multi-Family Rental Housing: From Trailer to Tower." *Journal of Urban Affairs* 42, 7.

August, Martine, and Alan Walks. 2018. "Gentrification, Suburban Decline, and the Financialization of Multi-Family Rental Housing: The Case of Toronto." *Geoforum* 89.

August, Martine, and Cole Webber. 2019. "Demanding the Right to the City and the Right to Housing (R2C/R2H): Best Practices for Supporting Community Organizing." Toronto: Parkdale Community Legal Services.

Baxter, Richard, and Katherine Brickell. 2014. "For Home UnMaking." *Home Cultures* 11, 2. https://doi.org/10.2752/175174214X13891916944553.

Beder, Sharon. 1995. "SLAPPS — Strategic Lawsuits Against Public Participation: Coming to a Controversy Near You." *Current Affairs Bulletin* 72, 3.

Bell, Danielle. 2013. "Shots Fired in Troubled 'Hood." *Ottawa Sun*, November 28: 16.

___. 2014a. "'People Are Afraid Here'; Rampant Crime Rattles Heron Gate." *Ottawa Sun*, October 8.

___. 2014b. "Residents Living in Fear; Herongate Community Plagued by Gun Violence." *Ottawa Sun*, July 4.

Bevington, Douglas, and Chris Dixon. 2005. "Movement-Relevant Theory: Rethinking

Social Movement Scholarship and Activism." *Social Movement Studies* 4, 3.

Bisaillon, Laura. 2012. "An Analytic Glossary to Social Inquiry Using Institutional and Political Activist Ethnography." *International Journal of Qualitative Methods* 11, 5. https://doi.org/10.1177/160940691201100506.

Bizzarri, Ugo. 2016. "South Ottawa – A Timbercreek Development." Timbercreek Asset Management presentation slides, Ottawa Real Estate Forum.

Blakeley, Grace. 2019. *Stolen: How to Save the World from Financialisation.* London: Repeater.

Blatman-Thomas, Naama, and Libby Porter. 2019. "Placing Property: Theorizing the Urban from Settler Colonial Cities." *International Journal of Urban and Regional Research* 43, 1.

Bledsoe, Adam, and Willie Jamaal Wright. 2018. "The Pluralities of Black Geographies. *Antipode* 51, 2. https://doi.org/10.1111/anti.12467.

Blomley, Nicholas. 2003. "Law, Property, and the Geography of Violence: The Frontier, the Survey, and the Grid." *Annals of the Association of American Geographers* 93, 1. https://doi.org/10.1111/1467-8306.93109.

___. 2004. *Unsettling the City: Urban Land and the Politics of Property.* New York and London: Routledge.

Boykoff, Jules. 2007. "Limiting Dissent: The Mechanisms of State Repression in the USA." *Social Movement Studies* 6, 3.

Brah, Avtar. 1996. *Cartographies of Diaspora: Contesting Identities.* London and New York: Routledge.

Brand, Dionne. 2001. *A Map to the Door of No Return: Notes to Belonging.* Toronto: Doubleday Canada.

Brent, Paul. 2016. "Multi-Res Prez Sees Growth Ahead for Timbercreek." *Renx.ca*, August 11. renx.ca/multi-res-prez-timbercreek-will-spend-compete.

Brickell, Katherine, Melissa Fernández Arrigoitia, and Alex Vasudevan. 2017. "Geographies of Forced Eviction: Dispossession, Violence, Resistance." In *Geographies of Forced Eviction: Dispossession, Violence, Resistance*, edited by Katherine Brickell, Melissa Fernández Arrigoitia, and Alexander Vasudevan. London: Palgrave Macmillan UK. https://doi.org/10.1057/978-1-137-51127-0_1.

Brown-Saracino, Japonica (ed.). 2013. *The Gentrification Debates: A Reader.* New York: Routledge.

Burns, Robyn, and Lisbeth A. Berbary. 2021. "Placemaking as Unmaking: Settler Colonialism, Gentrification, and the Myth of 'Revitalized' Urban Spaces." *Leisure Sciences* 43, 6. http://www.tandfonline.com/doi/abs/10.1080/01490400.2020.1870592.

Butler, Judith. 2004. *Precarious Life: The Powers of Mourning and Violence.* London: Verso.

___. 2009. *Frames of War: When Is Life Grievable?* London: Verso.

___. 2020. *The Force of Non-Violence.* London: Verso.

Canadian Apartment. 2015. "Top 10 in the Canadian Apartment Industry." July 27. reminetwork.com/articles/top-10-in-the-canadian-apartment-industry.

Canadian Real Estate Forums. 2019. "2019 Conference Program." realestateforums.com/caic/en/program/2019-program.html.

Canan, Penelope. 1989. "The SLAPP from a Sociological Perspective." *Pace Environmental Law Review* 7.

CAPREIT. 2018. "2018 Annual Report." Canadian Apartment Properties Real Estate

Investment Trust. annualreports.com/HostedData/AnnualReportArchive/C/TSX_CAR.UN_2018.pdf.

Cassidy, Robert. 1980. *Livable Cities: A Grass-Roots Guide to Rebuilding Urban America*. New York: Holt, Rinehart and Winston.

Cavanagh, Edward. 2011. "Review Essay: Discussing Settler Colonialism's Spatial Cultures." *Settler Colonial Studies* 1, 1.

CBC. 2012. "Trouble for Rent." *Marketplace* Season 39, Episode 3. gem.cbc.ca/media/marketplace/s39e03.

Chaddha, Anmol, and William Julius Wilson. 2008. "Reconsidering the 'Ghetto.'" *City & Community* 7, 4.

Chandler, Scott. 2017. "Multi-Residential Market and Investment Trends." Colliers International presentation slides, Canadian Apartment Investment Conference.

Chianello, Joanne, and Kat Porter. 2021a. Integrity Commissioner Recommends Coun. Jan Harder Be Removed as Planning Chair." *CBC News*, June 18. cbc.ca/news/canada/ottawa/ottawa-integrity-commissioner-report-1-6066736-1.6066736.

___. 2021b. "Jan Harder Resigns from Committee as Gloves Come Off at City Council." *CBC News*, June 23. cbc.ca/news/canada/ottawa/jan-harder-resigns-planning-committee-city-council-1.6076668.

Chisholm, Elinor, Philipa Howden-Chapman, and Geoff Fougere. 2020. "Tenants' Responses to Substandard Housing: Hidden and Invisible Power and the Failure of Rental Housing Regulation." *Housing, Theory and Society* 37, 2.

Chiu, Rebecca L.H. 2019. "Liveable Cities/Urban Liveability." In *The Wiley Blackwell Encyclopedia of Urban and Regional Studies*, edited by Anthony M. Orum. John Wiley & Sons. https://doi.org/10.1002/9781118568446.eurs0184.

Choudry, Aziz. 2013. "Activist Research Practice: Exploring Research and Knowledge Production for Social Action." *Socialist Studies* 9, 1. https://doi.org/10.18740/S4G01K.

___. 2014. "Activist Research for Education and Social Movement Mobilization." *Postcolonial Directions in Education* 3, 1.

___. 2015. *Learning Activism: The Intellectual Life of Contemporary Social Movements*. Toronto: University of Toronto Press.

___. 2019. "Social Movement Knowledge Production." In *Handbook of Theory and Research in Cultural Studies and Education*, edited by Peter Pericles Trifonas. Cham: Springer International Publishing. https://doi.org/10.1007/978-3-030-01426-1_59-1.

Christophers, Brett. 2020. *Rentier Capitalism: Who Owns the Economy, and Who Pays for It?* London: Verso.

City of Ottawa. 2018. "Heron Gate Secondary Plan Community Open House: Shaping the Vision for a Greater Community." January 23. jeancloutier.com/wp-content/uploads/2018/02/Jan-23-Heron-Gate-Open-House-PowerPoint-Slides.pdf.

___. 2019a. *5 Big Moves*. ehq-production-canada.s3.ca-central-1.amazonaws.com/documents/attachments/43b9dc66d91e7d1aeb6abd7731847355fadd49a0/000/018/759/original/OP_5BigMoves_EN_FINAL-for_posting.pdf.

___. 2019b. *Regeneration*. ehq-production-canada.s3.ca-central-1.amazonaws.com/7f48988d30792a899471e84710e8bcc6cd10dfbb/original/1605881262/Formatted_Regeneration_EW_FINAL.pdf_b88bbafb8ddb5a45c8530bfd2348ef2e.

___. 2021a. *New Official Plan: Report to Joint Meeting of Planning Committee and Agriculture and Rural Affairs Committee*. File Number ACS2021-PIE-EDP-0036.

___. 2021b. "Official Plan." ottawa.ca/en/planning-development-and-construction/

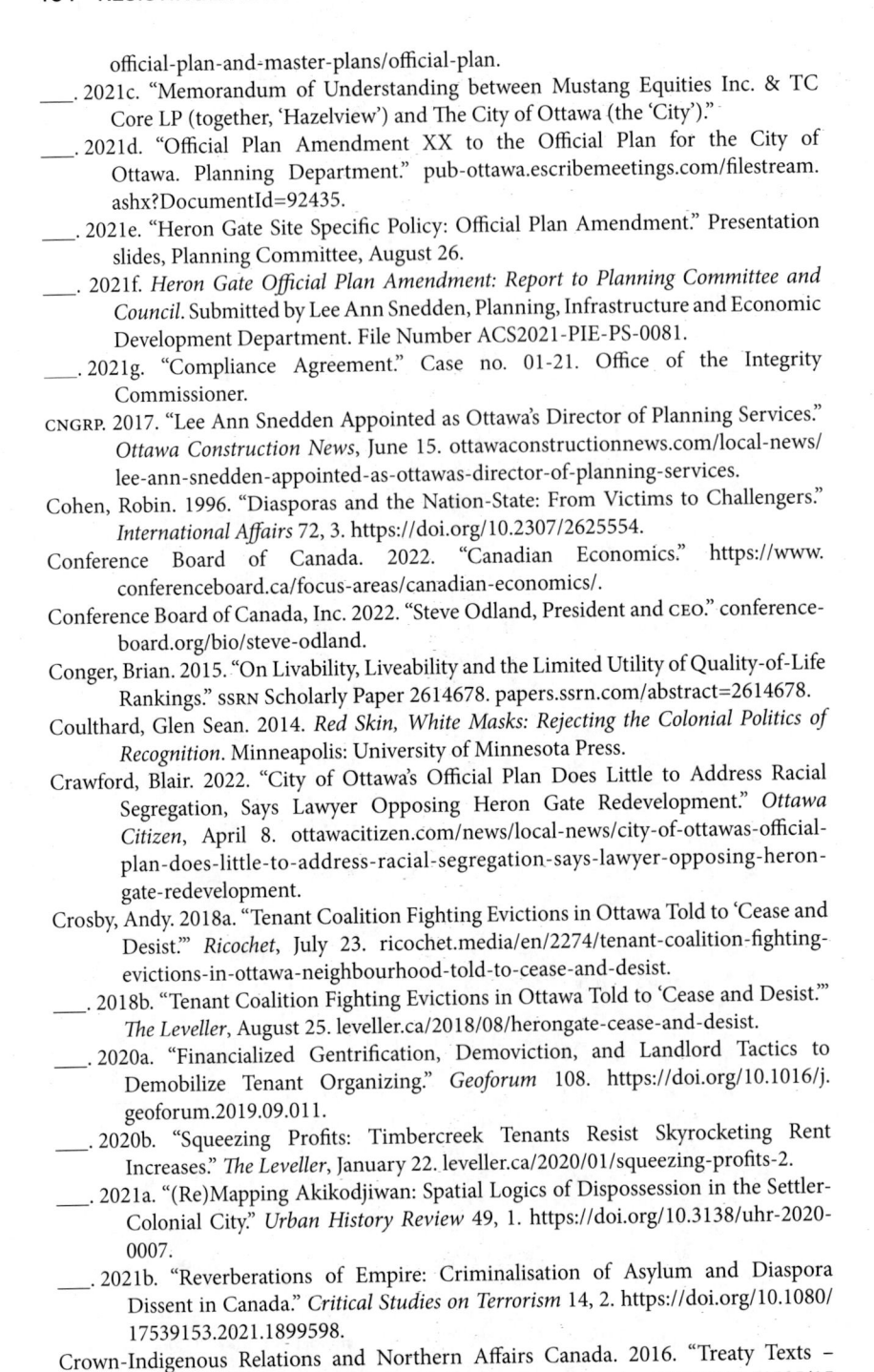

official-plan-and-master-plans/official-plan.
———. 2021c. "Memorandum of Understanding between Mustang Equities Inc. & TC Core LP (together, 'Hazelview') and The City of Ottawa (the 'City')."
———. 2021d. "Official Plan Amendment XX to the Official Plan for the City of Ottawa. Planning Department." pub-ottawa.escribemeetings.com/filestream.ashx?DocumentId=92435.
———. 2021e. "Heron Gate Site Specific Policy: Official Plan Amendment." Presentation slides, Planning Committee, August 26.
———. 2021f. *Heron Gate Official Plan Amendment: Report to Planning Committee and Council.* Submitted by Lee Ann Snedden, Planning, Infrastructure and Economic Development Department. File Number ACS2021-PIE-PS-0081.
———. 2021g. "Compliance Agreement." Case no. 01-21. Office of the Integrity Commissioner.
CNGRP. 2017. "Lee Ann Snedden Appointed as Ottawa's Director of Planning Services." *Ottawa Construction News*, June 15. ottawaconstructionnews.com/local-news/lee-ann-snedden-appointed-as-ottawas-director-of-planning-services.
Cohen, Robin. 1996. "Diasporas and the Nation-State: From Victims to Challengers." *International Affairs* 72, 3. https://doi.org/10.2307/2625554.
Conference Board of Canada. 2022. "Canadian Economics." https://www.conferenceboard.ca/focus-areas/canadian-economics/.
Conference Board of Canada, Inc. 2022. "Steve Odland, President and CEO." conference-board.org/bio/steve-odland.
Conger, Brian. 2015. "On Livability, Liveability and the Limited Utility of Quality-of-Life Rankings." SSRN Scholarly Paper 2614578. papers.ssrn.com/abstract=2614678.
Coulthard, Glen Sean. 2014. *Red Skin, White Masks: Rejecting the Colonial Politics of Recognition.* Minneapolis: University of Minnesota Press.
Crawford, Blair. 2022. "City of Ottawa's Official Plan Does Little to Address Racial Segregation, Says Lawyer Opposing Heron Gate Redevelopment." *Ottawa Citizen*, April 8. ottawacitizen.com/news/local-news/city-of-ottawas-official-plan-does-little-to-address-racial-segregation-says-lawyer-opposing-heron-gate-redevelopment.
Crosby, Andy. 2018a. "Tenant Coalition Fighting Evictions in Ottawa Told to 'Cease and Desist.'" *Ricochet*, July 23. ricochet.media/en/2274/tenant-coalition-fighting-evictions-in-ottawa-neighbourhood-told-to-cease-and-desist.
———. 2018b. "Tenant Coalition Fighting Evictions in Ottawa Told to 'Cease and Desist.'" *The Leveller*, August 25. leveller.ca/2018/08/herongate-cease-and-desist.
———. 2020a. "Financialized Gentrification, Demoviction, and Landlord Tactics to Demobilize Tenant Organizing." *Geoforum* 108. https://doi.org/10.1016/j.geoforum.2019.09.011.
———. 2020b. "Squeezing Profits: Timbercreek Tenants Resist Skyrocketing Rent Increases." *The Leveller*, January 22. leveller.ca/2020/01/squeezing-profits-2.
———. 2021a. "(Re)Mapping Akikodjiwan: Spatial Logics of Dispossession in the Settler-Colonial City." *Urban History Review* 49, 1. https://doi.org/10.3138/uhr-2020-0007.
———. 2021b. "Reverberations of Empire: Criminalisation of Asylum and Diaspora Dissent in Canada." *Critical Studies on Terrorism* 14, 2. https://doi.org/10.1080/17539153.2021.1899598.
Crown-Indigenous Relations and Northern Affairs Canada. 2016. "Treaty Texts – Upper Canada Land Surrenders." rcaanc-cirnac.gc.ca/eng/1370372152585/15

81293792285.

CTV *Ottawa*. 2016. "Ottawa Police Use Vacant Homes for Explosive Tactical Training." CTV *Ottawa*, March 22. ottawa.ctvnews.ca/news/latest-news/ottawa-police-use-vacant-homes-for-explosive-tactical-training-1.2828333.

Dantzler, Prentiss A. 2021. "The Urban Process under Racial Capitalism: Race, Anti-Blackness, and Capital Accumulation." *Journal of Race, Ethnicity and the City* 2, 2. https://doi.org/10.1080/26884674.2021.1934201.

Davenport, Christian. 2007. "State Repression and Political Order." *Annual Review of Political Science* 10.

Delamont, Kieran. 2019. "How the Fight for This Immigrant Neighbourhood Became a Fight for All Immigrant Neighbourhoods." *TVO*, August 29. tvo.org/article/how-the-fight-for-this-immigrant-neighbourhood-became-a-fight-for-all-immigrant-neighbourhoods.

Desmond, Matthew. 2016. *Evicted: Poverty and Profit in the American City*. New York: Broadway Books.

Devault, Marjorie L. 2006. "Introduction: What Is Institutional Ethnography?" *Social Problems* 53, 3. https://doi.org/doi:10.1525/sp.2006.53.3.294.

Devault, Marjorie L., and Liza McCoy. 2001. "Institutional Ethnography: Using Interviews to Investigate Ruling Relations." In *Handbook of Interview Research*, edited by Jaber F. Gubrium and James A. Holstein. Thousand Oaks, CA: Sage Publications. https://doi.org/10.4135/9781412973588.n43.

DIALOG. 2022. "Community Wellbeing Framework." dialogdesign.ca/community-wellbeing-framework.

Diwan, Faryal, William Turman, Drew Baird, Neelu Mehta, Aleksandra Petrovic, and Brian Doucet. 2021. "Mapping Displacement in Kitchener-Waterloo: Report." Kitchener, ON: Social Development Centre Waterloo Region.

Dixon, Chris. 2014. *Another Politics: Talking Across Today's Transformative Movements*. Oakland: University of California Press.

Domaradzka, Anna. 2019. "The Un-Equal Playground: Developers and Urban Activists Struggling for the Right to the City." *Geoforum* 134. https://doi.org/10.1016/j.geoforum.2019.01.013.

Donnan, Mary Ellen. 2016. "Domicide and Indigenous Homelessness in Canada." *Journal of Sociology and Social Work* 4, 2. https://doi.org/10.15640/jssw.v4n2a5.

Donson, Fiona J.L. 2000. *Legal Intimidation: A SLAPP in the Face of Democracy*. London: Free Association Books.

Dorries, Heather, Robert Henry, David Hugill, Tyler McCreary, and Julie Tomiak (eds.). 2019. *Settler City Limits: Indigenous Resurgence and Colonial Violence in the Urban Prairies*. Winnipeg: University of Manitoba Press.

Dorries, Heather, David Hugill, and Julie Tomiak. 2019. "Racial Capitalism and the Production of Settler Colonial Cities." *Geoforum* 132. https://doi.org/10.1016/j.geoforum.2019.07.016.

Earl, Jennifer. 2011. "Political Repression: Iron Fists, Velvet Gloves, and Diffuse Control." *Annual Review of Sociology* 37. https://doi.org/10.1146/annurev.soc.012809.102609.

Edmonds, Penelope. 2010. *Urbanizing Frontiers: Indigenous Peoples and Settlers in 19th-Century Pacific Rim Cities*. Vancouver: University of British Columbia Press.

Egal, Mumina. 2018. "Muslims Facing Eviction in Ottawa Neighbourhood Hope You Will Stand Up for Them and for Affordable Housing." *Muslim Link*, May 31. muslimlink.ca/news/muslims-facing-eviction-ottawa-herongate-timbercreek-affordable-housing.

Egal, Mumina, and Josh Hawley. 2018. "From Hard Numbers to Harder Reality." *The Leveller*, November 14. leveller.ca/2018/11/from-hard-numbers-to-harder-reality.

Ehrenfeucht, Renia, and Marla Nelson. 2020. "Just Revitalization in Shrinking and Shrunken Cities? Observations on Gentrification from New Orleans and Cincinnati." *Journal of Urban Affairs* 42, 3.

Ellefsen, Rune. 2016. "Judicial Opportunities and the Death of SHAC: Legal Repression along a Cycle of Contention." *Social Movement Studies* 15, 5.

Elliott-Cooper, Adam, Phil Hubbard, and Loretta Lees. 2020. "Moving beyond Marcuse: Gentrification, Displacement and the Violence of Un-Homing." *Progress in Human Geography* 44, 3.

Ellis-Young, Margaret. 2022. "Gentrification as (Settler) Colonialism? Moving beyond Metaphorical Linkages." *Geography Compass* 16, 1. https://doi.org/10.1111/gec3.12604.

Farha, Leilani. 2018. "We Can't Turn a Blind Eye to Investment Firms Preying on Low-Income Homes." *The Huffington Post*, July 5. huffingtonpost.ca/leilani-farha/evict-fix-up-cash-out-companies-prey-on-low-income-homes-for-profits_a_23475563.

Fawcett, R. Ben, and Ryan Walker. 2020. "Indigenous Peoples, Indigenous Cities." In *Canadian Cities in Transition: Understanding Contemporary Urbanism*, edited by Markus Moos, Tara Vinodrai, and Ryan Walker. 6th ed. Don Mills, ON: Oxford University Press.

Fields, Desiree. 2015. "Contesting the Financialization of Urban Space: Community Organizations and the Struggle to Preserve Affordable Rental Housing in New York City." *Journal of Urban Affairs* 37, 2. https://doi.org/10.1111/juaf.12098.

___. 2017. "Unwilling Subjects of Financialization." *International Journal of Urban and Regional Research* 41, 4. https://doi.org/10.1111/1468-2427.12519.

Fields, Desiree, and Elora Lee Raymond. 2021. "Racialized Geographies of Housing Financialization." *Progress in Human Geography* 45, 6. https://doi.org/10.1177/03091325211009299.

Fields, Desiree, and Sabina Uffer. 2016. "The Financialisation of Rental Housing: A Comparative Analysis of New York City and Berlin." *Urban Studies* 53, 7.

Fox, Carl. 2020. "Free Speech, Public Shaming, and the Role of Social Media." In *Smart Technologies and Fundamental Rights*, edited by John-Stewart Gordon. Leiden, Netherlands: Brill.

Frampton, Caelie, Gary Kinsman, A.K. Thompson, and Kate Tilleczek. 2006. "Social Movements/Social Research: Towards Political Activist Ethnography." In *Sociology for Changing the World: Social Movements/Social Research*, edited by Caelie Frampton, Gary Kinsman, A.K. Thompson, and Kate Tilleczek. Halifax, NS: Fernwood Publishing.

Freeman, Lance. 2006. *There goes the Hood: Views of Gentrification from the Ground Up*. Temple University Press.

Freeman, Lance, and Frank Braconi. 2004. "Gentrification and Displacement: New York City in the 1990s." *Journal of the American Planning Association* 70, 1.

Fullilove, Mindy Thompson. 2016. *Root Shock: How Tearing Up City Neighbourhoods Hurts America, and what we can do about it*. New Village Press.

Gagné, Natacha, and Benoît Trépied. 2016. "Introduction to Special Issue Colonialism, Law, and the City: The Politics of Urban Indigeneity." *City & Society* 28, 1.

Gamson, William A. 2009. "Defining Movement 'Success'" In *The Social Movements Reader: Cases and Concepts*, edited by Jeff Goodwin and James M. Jasper. 2nd ed. Malden, MA: Wiley-Blackwell.

Gashan and Ali v. Timbercreek Asset Management et al. 2019. Court File No. 19-SC-153021 (Ontario Superior Court of Justice).

Gehl, Lynn. 2014. *The Truth that Wampum Tells: My Debwewin on the Algonquin Land Claims Process*. Halifax, NS: Fernwood Publishing.

Gilmore, Ruth Wilson. 2007. *Golden Gulag: Prisons, Surplus, Crisis, and Opposition in Globalizing California*. Berkeley and Los Angeles, CA: University of California Press.

Government of Ontario. 2020. "Provincial Policy Statement, 2020: Under the Planning Act." Order in Council No. 229/2020.

Guenther, Lisa. 2019. "Seeing Like a Cop: A Critical Phenomenology of Whiteness. In *Race as Phenomena: Between Phenomenology and Philosophy of Race*, edited by Emily S. Lee. Lanham, MD: Rowman & Littlefield.

Hall, Stuart. 1990. "Cultural identity and Diaspora." In *Identity: Community, Culture, Difference*, edited by Jonathan Rutherford. London: Lawrence & Wishart.

Halpin, Bryony Jane. 2017. "Unsettling Revitalization in Toronto: The Fantasy and Apology of the Settler City." PhD diss., York University.

Hankins, Katherine B., and Emily M. Powers. 2009. "The Disappearance of the State from 'Livable' Urban Spaces." *Antipode* 41, 5. https://doi.org/10.1111/j.1467-8330.2009.00699.x.

Harris, Cole. 2004. "How did Colonialism Dispossess? Comments from an Edge of Empire." *Annals of the Association of American Geographers* 94, 1.

___. 2020. *A Bounded Land: Reflections on Settler Colonialism in Canada*. Vancouver: University of British Columbia Press.

Harvey, David. 1974. Class-monopoly Rent, Finance Capital and the Urban Revolution. *Regional Studies* 8, 3–4. https://doi.org/10.1080/09595237400185251.

___. 1978. The Urban Process under Capitalism: A Framework for Analysis. *International Journal of Urban and Regional Research* 2, 1–3. https://doi.org/10.1111/j.1468-2427.1978.tb00738.x.

Hawley, Josh. 2018. "The Money behind the Heron Gate Evictions." *The Leveller*, September 21. leveller.ca/2018/09/the-german-money-behind-the-heron-gate-evictions.

Hawthorne, Camilla. 2019. "Black Matters are Spatial Matters: Black Geographies for the Twenty-first Century." *Geography Compass* 13, 11. https://doi.org/10.1111/gec3.12468.

Haynes, Bruce, and Ray Hutchison. 2008. "The Ghetto: Origins, History, Discourse." *City & Community* 7, 4. https://doi.org/10.1111/j.1540-6040.2008.00271_1.x.

Hazelview Investments. 2022. "About Us." hazelview.com/about.

Hazelview Properties. 2021. "Vista Local." hazelviewproperties.com/residential/vista-local.

Henig, Jeffrey R. 1980. "Gentrification and Displacement within Cities: A Comparative Analysis." *Social Science Quarterly* 61, 3–4.

Herongate Tenant Coalition. 2018a. "The battle for Heron Gate." *Briarpatch* 47, 5.

___. 2018b. "Support Herongate Tenants." herongatetenants.ca/we-are-herongate-issue-3/support-herongate-tenants.

___. 2018c. *Herongate Tenant Coalition – Tenants' Rights Meeting*. Video. youtube.com/watch?v=cJKUDkxVX94.

___. 2018d. "About." herongatetenants.ca/about/.
___. 2018e. *Notice to Tenants: Here Is What You Need to Know.*
___. 2018f. "Herongate Eviction Census." herongatetenants.ca/herongate-eviction-census.
___. 2018g. "Severe Inequality in Canada's Capital." herongatetenants.ca/our-neighbourhood/severe-inequality-in-canadas-capital.
___. 2020. *My Community is Here – Herongate Mass Evictions.* YouTube video, January 22. youtube.com/watch?v=uJKLv-VYNJM.
___. 2021a. "The Bombing of Herongate." herongatetenants.ca/our-neighbourhood/the-bombing-of-herongate.
___. 2021b. "A Framework for Destruction: Hazelview's Plans to Demolish 559 More Homes and Eliminate Affordable Housing." Presentation, November 21.
Hilson, Christopher J. 2016 "Environmental SLAPPs in the UK: Threat or Opportunity? *Environmental Politics* 25, 2.
Horgan, Mervyn. 2018. "Territorial Stigmatization and Territorial Destigmatization: A Cultural Sociology of Symbolic Strategy in the Gentrification of Parkdale (Toronto)." *International Journal of Urban and Regional Research* 42, 3. https://doi.org/10.1111/1468-2427.12645.
Hugill, David. 2017. "What is a Settler-Colonial City?" *Geography Compass* 11, 5.
___. 2019. "Comparative Settler Colonial Urbanisms: Racism and the Making of Inner-City Winnipeg and Minneapolis, 1940-1975." In *Settler City Limits: Indigenous Resurgence and Colonial Violence in the Urban Prairies*, edited by Heather Dorries, Robert Henry, David Hugill, Tyler McCreary, and Julie Tomiak. Winnipeg: University of Manitoba Press.
Huitema, Marijke, Brian S. Osborne, and Michael Ripmeester. 2002. "Imagined Spaces, Constructed Boundaries, Conflicting Claims: A Legacy of Postcolonial Conflict in Eastern Ontario." *International Journal of Canadian Studies* 25.
Human Rights Council. 2017. *Report of the Special Rapporteur on Adequate Housing as a Component of the Right to an Adequate Standard of Living, and on the Right to Non-discrimination in this Context.* United Nations General Assembly, A/HRC/34/51.
Hussein, Nima, and Josh Hawley. 2021. "Uneven Development, Discrimination in Housing and Organized Resistance." In *Beyond Free Market: Social Inclusion and Globalization*, edited by Fayyaz Baqir and Sanni Yaya. London and New York: Routledge.
Hyde, Zachary. 2022. "Giving Back to Get Ahead: Altruism as a Developer Strategy of Accumulation through Affordable Housing Policy in Toronto and Vancouver." *Geoforum* 134. https://doi.org/10.1016/j.geoforum.2018.07.005.
Hyra, Derek S. 2012. "Conceptualizing the New Urban Renewal: Comparing the Past to the Present." *Urban Affairs Review* 48, 4. https://doi.org/10.1177/1078087411434905.
Hyra, Derek, Dominic Moulden, Carley Weted, and Mindy Fullilove. 2019. "A Method for Making the Just City: Housing, Gentrification, and Health." *Housing Policy Debate* 29, 3. https://doi.org/10.1080/10511482.2018.1529695.
InterRent REIT. 2014. *2014 Annual Report.*
IPE Real Assets. 2015. "Canada: Exporting Expertise." realassets.ipe.com/markets-/regions/americas/canada-exporting-expertise/10007046.article.
Iyanda, Ayodeji Emmanuel, and Yongmei Lu. 2021. "Perceived Impact of Gentrification on Health and Well-being: Exploring Social Capital and Coping Strategies in

Gentrifying Neighborhoods." *The Professional Geographer* 73, 4.
Jacobs, Jane. 1961. *The Death and Life of Great American Cities*. New York: Random House.
James, Ryan K. 2010. "From 'Slum Clearance' to 'Revitalisation': Planning, Expertise and Moral Regulation in Toronto's Regent Park." *Planning Perspectives* 25, 1.
Jelks, Na'Taki Osborne, Viniece Jennings, and Alessandro Rigolon. 2021. "Green Gentrification and Health: A Scoping Review." *International Journal of Environmental Research and Public Health* 18, 3.
Jiwani, Yasmin, and Ahmed Al-Rawi. 2021. "Intersecting Violence: Representations of Somali Youth in the Canadian Press." *Journalism* 22, 7.
Jokic, Dallas. 2020. "Cultivating the Soil of White Nationalism: Settler Violence and Whiteness as Territory." *Journal of Critical Race Inquiry* 7, 2.
Kaal, Harm. 2011. "A Conceptual History of Livability: Dutch Scientists, Politicians, Policy makers and Citizens and the Quest for a Livable City." *City* 15, 5.
Kadıoğlu, Defne. 2022. "Producing Gentrifiable Neighborhoods: Race, Stigma and Struggle in Berlin-Neukölln." *Housing Studies*. https://doi.org/10.1080/02673037.2022.2042494.
Kent-Stoll, Peter. 2020. "The Racial and Colonial Dimensions of Gentrification." *Sociology Compass* 14, 12. https://doi.org/10.1111/soc4.12838.
Kinsman, Gary. 2006. "Mapping Social Relations of Struggle: Activism, Ethnography, Social Organization." In *Sociology for Changing the World*, edited by Caelie Frampton, Gary Kinsman, A.K. Thompson, and Kate Tilleczek. Halifax, NS: Fernwood Publishing.
___. 2023. "Direct Action as Political Activist Ethnography: Activist Research in the Sudbury Coalition Against Poverty." In *Political Activist Ethnography: Studies in the Social Relations of Struggle*, edited by Agnieszka Doll, Laura Bisaillon, and Kevin Walby. Athabasca, AB: Athabasca University Press.
Komakech, Morris D.C., and Suzanne F. Jackson. 2016. "A Study of the Role of Small Ethnic Retail Grocery Stores in Urban Renewal in a Social Housing Project, Toronto, Canada." *Journal of Urban Health* 93. https://doi.org/10.1007/s11524-016-0041-1.
Kusow, Abdi M. 2001. "Stigma and Social Identities: The Process of Identity Work among Somali Immigrants in Canada." In *Variations on the Theme of Somaliness*, edited by Muddle Suzanne Lilius. Centre for Continuing Education, Åbo Akademi University.
Landry, Normand. 2014. *Threatening Democracy: SLAPPs and the Judicial Repression of Political Discourse*. Halifax, NS: Fernwood Publishing.
Launius, Sarah, and Geoffrey Alan Boyce. 2021. "More than Metaphor: Settler Colonialism, Frontier Logic, and the Continuities of Racialized Dispossession in a Southwest U.S. City." *Annals of the American Association of Geographers* 111, 1. https://doi.org/10.1080/24694452.2020.1750940.
Lawrence, Bonita. 2012. *Fractured Homeland: Federal Recognition and Algonquin Identity in Ontario*. Vancouver: University of British Columbia Press.
Leach, Joanne M., Peter A. Braithwaite, Susan E. Lee, Christopher J. Bouch, Dexter V.L. Hunt, and Chris D.F. Rogers. 2016. "Measuring Urban Sustainability and Liveability Performance: The City Analysis Methodology." *International Journal of Complexity in Applied Science and Technology* 1, 1. https://doi.org/10.1504/IJCAST.2016.081296.
Lewis, Nemoy. 2022. The Uneven Racialized Impacts of Financialization: A Report for the

Office of the Federal Housing Advocate. Canadian Human Rights Commission.

Ley, David. 1990. "Urban Liveability in Context." *Urban Geography* 11, 1. https://doi.org/10.2747/0272-3638.11.1.31.

Link2Build Ontario. 2021. "South Ottawa Development OK to Proceed." September 10. link2build.ca/news/articles/2021/september/south-ottawa-development-ok-to-proceed/.

Lord, Craig. 2021. "Ottawa Planning Chair Faces Calls for Resignation after Integrity Commissioner Report." *Global News*, June 21. globalnews.ca/news/7968684/ottawa-planning-chair-integrity-report-jan-harder/.

Loury, Glenn C. 2021. *The Anatomy of Racial Inequality, With a New Preface*. Harvard University Press.

Lowe, Melanie, Carolyn Whitzman, Hannah Badland, Melanie Davern, Dominque Hes, Lu Aye, Iain Butterworth, and Billie Giles-Corti. 2013. *Liveable, Healthy, Sustainable: What Are the Key Indicators for Melbourne Neighbourhoods?* Melbourne: University of Melbourne.

Lub, Vasco. 2018. *Neighbourhood Watch in a Digital Age: Between Crime Control and Culture of Control*. Cham, Switzerland: Springer.

Lubbers, Eveline. 2015. "Undercover Research: Corporate and Police Spying on Activists. An Introduction to Activist Intelligence as a New Field of Study." *Surveillance and Society* 13, 3–4.

Macionis, John J., and Vincent N. Parrillo. 2016. *Cities and Urban Life*. 7th ed. Boston: Pearson.

Madden, David, and Peter Marcuse. 2016. *In Defense of Housing: The Politics of Crisis*. London: Verso.

Majaury, Heather. 2005. "Living Inside Layers of Colonial Division: A Part of the Algonquin Story." *Atlantis* 29, 2.

Mar, Tracey Banivanua, and Penelope Edmonds. 2010. "Introduction: Making Space in Settler Colonies." In *Making Settler Colonial Space*, edited by Tracey Banivanua Mar and Penelope Edmonds. London: Palgrave Macmillan.

Markovich, Julia, Monika Slovinec D'Angelo, and Thy Dinh. 2018. *Community Wellbeing: A Framework for the Design Professions*. Ottawa: The Conference Board of Canada.

Mast, Tammy, and Josh Hawley. 2021. "A Framework for Destruction: Hazelview and the City's Plan to Demolish 559 More Homes in Heron Gate Village." *The Leveller*, December 20. leveller.ca/2021/12/a-framework-for-destruction/.

Masuda, Jeffrey R., and Sonia Bookman. 2018. "Neighbourhood Branding and the Right to the City." *Progress in Human Geography* 42, 2.

Mayblin, Lucy. 2017. *Asylum after Empire: Colonial Legacies in the Politics of Asylum Seeking*. London: Rowman & Littlefield Publishers.

Mayer, Margit. 2020. "What Does It Mean to Be a (Radical) Urban Scholar-Activist, or Activist Scholar, Today?" *City* 24, 1–2. https://doi.org/10.1080/13604813.2020.1739909.

Maynard, Robyn. 2017. *Policing Black Lives: State Violence in Canada from Slavery to the Present*. Halifax & Winnipeg: Fernwood Publishing.

McCann, Eugene. 2004. "'Best Places': Interurban Competition, Quality of Life and Popular Media Discourse." *Urban Studies* 41.

———. 2007. "Inequality and Politics in the Creative City-Region: Questions of Livability and State Strategy." *International Journal of Urban and Regional Research* 31, 1. https://doi.org/10.1111/j.1468-2427.2007.00713.x.

McClintock, Nathan. 2018. "Urban Agriculture, Racial Capitalism, and Resistance in the Settler-Colonial City." *Geography Compass* 12, 6.
McConville, Mike, and Dan Shepherd. 1992. *Watching Police, Watching Communities*. London: Routledge.
McCracken, Erin. 2015. "Agencies Rally to Support Evicted Herongate Residents." *Toronto.com*, October 15. bramptonguardian.com/news-story/5961586-agencies-rally-to-support-evicted-herongate-residents/.
___. 2016a. "'Three Monsters' Complex Draws Criticism." Ottawa South *News*, September 29. issuu.com/emcsouth/docs/ottawasouthmanoticknews100616.
___. 2016b. "Police Tactical Unit Sets Up Shop in Vacant Herongate Homes." *Ottawa South News*, March 22.
___. 2017. "Petition Circulating to Challenge Proposed Herongate Development." *Ottawa South News*, February 21. durhamregion.com/news-story/7150744-petition-circulating-to-challenge-proposed-herongate-development/.
McGuirk, Pauline, and Robyn Dowling. 2011. "Governing Social Reproduction in Masterplanned Estates: Urban Politics and Everyday Life in Sydney." *Urban Studies* 48, 12. https://doi.org/10.1177/0042098011411950.
McKittrick, Katherine. 2006. *Demonic Grounds: Black Women and the Cartographies of Struggle*. University of Minnesota Press.
___. 2011. "On Plantations, Prisons, and a Black Sense of Place." *Social & Cultural Geography* 12, 8. https://doi.org/10.1080/14649365.2011.624280.
___. 2013. "Plantation Futures." *Small Axe: A Caribbean Journal of Criticism* 17, 3 (42) https://doi.org/10.1215/07990537-2378892.
McLean, Steve. 2017. "Residential Leaders Share Multi-Family Investment Strategies." *Renx.ca*, September 14. renx.ca/residential-multi-family-investment-strategies.
McNeilly, Kathryn. 2016. "Livability: Notes on the Thought of Judith Butler." *Critical Legal Thinking* (blog). May 26. criticallegalthinking.com/2016/05/26/livability-judith-butler/.
Melamed, Jodi. 2015. "Racial Capitalism." *Critical Ethnic Studies* 1, 1. https://doi.org/10.5749/jcritethnstud.1.1.0076.
Mensah, Joseph. 2019. *Expert Opinion on Discrimination Based on Race and Related Grounds in Rental Housing—Heron Gate Community, Ottawa, Canada*. Avant Law, LLP.
Mensah, Joseph, and Daniel Tucker-Simmons. 2021. "Social (In)Justice and Rental Housing Discrimination in Urban Canada: The Case of Ethno-Racial Minorities in the Herongate Community in Ottawa." *Studies in Social Justice* 15, 1. https://doi.org/10.26522/ssj.v15i1.2239.
Mensah, Joseph, and Christopher J. Williams. 2017. *Boomerang Ethics: How Racism Affects Us All*. Halifax, NS and Winnipeg, MB: Fernwood Publishing.
Meredith, Colin, and Chantal Paquette. 1992. "Crime Prevention in High-Rise Rental Apartments: Findings of a Demonstration Project." *Security Journal* 3, 3.
Meyer, David S. 2003. "How Social Movements Matter." *Contexts* 2, 4. https://doi.org/10.1525/ctx.2003.2.4.30.
Mielczarek, Natalia. 2018. "The 'Pepper-Spraying Cop' Icon and Its Internet Memes: Social Justice and Public Shaming through Rhetorical Transformation in Digital Culture." *Visual Communication Quarterly* 25, 2.
Miller, Jessica Ty. 2020. "Temporal Analysis of Displacement: Racial Capitalism and Settler Colonial Urban Space." *Geoforum* 116.
Miller, Peter, and Nikolas Rose. 2008. *Governing the Present: Administering Economic,*

Social and Personal Life. Cambridge, UK: Polity Press.
Minto. 2021. "About." minto.com/about-minto/index.html.
Momentum Planning & Communications. 2021. "Timbercreek Communities Ottawa and Toronto." momentumplancom.ca/our-work/timbercreek-communities-heron-gate-redevelopment-project/.
Montalva Barba, Miguel A. 2021. "(Re)enacting Settler Colonialism via White Resident Utterances." *Critical Sociology* 47, 7–8. https://journals.sagepub.com/doi/10.1177/0896920520976788.
Moores, Chris. 2017. "Thatcher's Troops? Neighbourhood Watch Schemes and the Search for 'Ordinary' Thatcherism in 1980s Britain." *Contemporary British History* 31, 2. https://doi.org/10.1080/13619462.2017.1306203.
Moreton-Robinson, Aileen. 2018. "White Possession and Indigenous Sovereignty Matters." In *Race and Racialization: Essential Readings, second edition*, edited by Tania Das Gupta, Carl E. James, Chris Andersen, Grace-Edward Galabuzi, and Roger C.A. Maaka. Toronto, ON: Canadian Scholars.
Morrison, James. 2005. *Algonquin History in the Ottawa River Watershed.* Ottawa: Sicani Research & Advisory Services.
Mosselson, Aidan. 2020. "Habitus, Spatial Capital and Making Place: Housing Developers and the Spatial Praxis of Johannesburg's Inner-City Regeneration." *Environment and Planning A: Economy and Space* 52, 2. https://doi.org/10.1177/0308518X19830970.
Mykhalovskiy, Eric. 2018. "Institutional Ethnography and Activist Futures." In *Frontiers of Global Sociology: Research Perspectives for the 21st Century*, edited by Markus S. Schulz. Berlin/New York: International Sociological Association Research.
Naegler, Laura. 2012. *Gentrification and Resistance: Cultural Criminology, Control, and the Commodification of Urban Protest in Hamburg.* Vol. 50. Münster: lit Verlag Münster.
Nagra, Baljit. 2017. *Securitized Citizens: Canadian Muslims' Experiences of Race Relations and Identity Formation Post–9/11.* Toronto, ON: University of Toronto Press.
Newman, Kathe, and Elvin Wyly. 2006. "The Right to Stay Put, Revisited: Gentrification and Resistance to Displacement in New York City." *Urban Studies* 43, 1.
Nowicki, Mel. 2017. "Domicide and the Coalition: Austerity, Citizenship and Moralities of Forced Eviction in Inner London." In *Geographies of Forced Eviction: Dispossession, Violence, Resistance*, edited by Katherine Brickell, Melissa Fernández Arrigoitia, and Alexander Vasudevan. London: Palgrave Macmillan UK. https://doi.org/10.1057/978-1-137-51127-0_6.
O'brien, Frank. 2018. "Smart REITS Are Playing the Land Development Card." *Western Investor*, February 27. westerninvestor.com/british-columbia/smart-reits-are-playing-the-land-development-card-3830335.
O'Reilly, Karen. 2011. *Ethnographic Methods.* 2nd ed. London: Routledge. https://doi.org/10.4324/9780203864722.
Onursal, Recep, and Daniel Kirkpatrick. 2021. "Is Extremism the 'New' Terrorism? The Convergence of 'Extremism' and 'Terrorism' in British Parliamentary Discourse." *Terrorism and Political Violence* 33, 5. https://doi.org/10.1080/09546553.2019.1598391.
Ooi, Giok Ling, and Belinda Yuen. 2009. *World Cities: Achieving Liveability and Vibrancy.* London: World Scientific. https://doi.org/10.1142/7398.
Osman, Laura. 2018. "Zibi's 'Pioneers' Take up Residence." *CBC News*, November 29. cbc.ca/news/canada/ottawa/zibi-condo-move-in-1.4923814.

Osman, Laura, and Joanne Chianello. 2018. "Developers Holding Fundraiser for Councillor on Planning Committee." *CBC News*, September 19. cbc.ca/news/canada/ottawa/fundraiser-planning-committee-developers-1.4829928.

Osterweil, Vicky. 2020. *In Defense of Looting: A Riotous History of Uncivil Action*. New York: Bold Type Books.

Ottawa Neighbourhood Study. 2021. "Ledbury – Heron Gate - Ridgemont." neighbourhoodstudy.ca/932ledbury-heron-gate-ridgemont.

Pacione, Michael. 1990. "Urban Liveability: A Review." *Urban Geography* 11, 1. https://doi.org/10.2747/0272-3638.11.1.1.

Palen, J. John, and Bruce London (eds.). 1984. *Gentrification, Displacement, and Neighborhood Revitalization*. New York: SUNY Press.

Parish, Jessica. 2020. "Re-Wilding Parkdale? Environmental Gentrification, Settler Colonialism, and the Reconfiguration of Nature in 21st Century Toronto." *Environment and Planning E: Nature and Space* 3, 1. https://doi.org/10.1177/2514848619868110.

Parker, Simon. 2004. *Urban Theory and the Urban Experience: Encountering the City*. London; New York: Routledge.

Pasternak, Shiri. 2017. *Grounded Authority: The Algonquins of Barriere Lake Against the State*. Minneapolis: University of Minnesota Press.

Payne, Elizabeth. 2019. "Lawsuit Alleges Timbercreek Failed to Act on Repeated Maintenance Requests." *Ottawa Citizen*, March 26. ottawacitizen.com/news/local-news/lawsuit-alleges-timbercreek-failed-to-act-on-repeated-maintenance-requests.

Pearson, Matthew. 2015. "Developers to Get Ambassadors Within Planning Department." *Ottawa Citizen*, June 1. ottawacitizen.com/news/local-news/developers-to-get-ambassadors-within-planning-department.

Picton, Roger M. 2010. "Selling National Urban Renewal: The National Film Board, the National Capital Commission and Post-War Planning in Ottawa, Canada." *Urban History* 37, 2. https://doi.org/10.1017/S0963926810000374.

Pilon, Jean-Luc, and Randy Boswell. 2015. "Below the Falls: An Ancient Cultural Landscape in the Centre of (Canada's National Capital Region) Gatineau." *Canadian Journal of Archaeology* 39.

Porteous, Douglas, and Sandra E. Smith. 2001. *Domicide: The Global Destruction of Home*. Montreal-Kingston: McGill-Queen's University Press.

Porter, Libby, and Oren Yiftachel. 2019. "Urbanizing Settler-Colonial Studies: Introduction to the Special Issue." *Settler Colonial Studies* 9, 2.

Power, Emily, and Bjarke Skærlund Risager. 2019. "Rent-Striking the REIT." *Radical Housing Journal* 1, 2.

Pring, George William, and Penelope Canan. 1996. *SLAPPs: Getting Sued for Speaking Out*. Philadelphia: Temple University Press.

Pullés, Stephanie A, and Jennifer Lee. 2019. "Ethnic Enclave." In *The Wiley Blackwell Encyclopedia of Urban and Regional Studies*, edited by Anthony M. Orum. John Wiley & Sons. https://doi.org/10.1002/9781118568446.eurs0092.

Quijano, Anibal. 2000. "Coloniality of Power and Eurocentrism in Latin America." *International Sociology* 15, 2. https://doi.org/10.1177/0268580900015002005.

Quizar, Jessi. 2019. "Land of Opportunity: Anti-Black and Settler Logics in the Gentrification of Detroit." *American Indian Culture and Research Journal* 43, 2. https://doi.org/10.17953/aicrj.43.2.quizar.

Randall, Steve. 2021. "More Than Half of Canadians Plan to Own Shares by Year-End."

Wealth Professional, September 3. wealthprofessional.ca/news/industry-news/more-than-half-of-canadians-plan-to-own-shares-by-year-end/359479.

Razack, Sherene. 1998. "Race, Space, and Prostitution: The Making of the Bourgeois Subject." *Canadian Journal of Women and the Law* 10, 2.

———. 2000. "Gendered Racial Violence and Spatialized Justice: The Murder of Pamela George." *Canadian Journal of Law and Society* 15, 2.

———. 2002. *Race, Space, and the Law: Unmapping a White Settler Society.* Toronto, ON: Between the Lines.

Reimer, Gwen. 2019. "British-Canada's Land Purchases, 1783-1788: A Strategic Perspective." *Ontario History* 111, 1. https://doi.org/10.7202/1059965ar.

Residential Tenancies Act, 2006, S.O. 2006, c.17.

Richardson, Boyce. 1993. *People of Terra Nullius: Betrayal and Rebirth in Aboriginal Canada.* Madeira Park, BC: Douglas & McIntyre.

Risager, Bjarke Skærlund. 2021. "Financialized Gentrification and Class Composition in the Post-Industrial City: A Rent Strike Against a Real Estate Investment Trust in Hamilton, Ontario." *International Journal of Urban and Regional Research* 45, 2. https://doi.org/10.1111/1468-2427.12991.

Robin, Enora. 2018. "Performing Real Estate Value(s): Real Estate Developers, Systems of Expertise and the Production of Space." *Geoforum* 134. https://doi.org/10.1016/j.geoforum.2018.05.006.

Robinson, Cedric. 2000. *Black Marxism: The Making of the Black Radical Tradition.* Chapel Hill, NC: University of North Carolina Press.

Rockwell, Neal. 2018a. "Patterns of Neglect." *The Leveller* 11, 3. https://leveller.ca/2018/11/patterns-of-neglect/.

———. 2018b. "Disconnected Realities: Country Club Politics Put Vulnerable Herongate Residents in Peril." *The Leveller*, September 21. leveller.ca/2018/09/disconnected-realities.

———. 2019. "Herongate Residents Contend with Below-Grade Conditions and a Rent Increase, While Councillors and Developers Mingle." *The Leveller*, March 21. leveller.ca/2019/03/below-grade-herongate-residents-contend-with-broken-heating-broken-pipes-and-a-rent-increase-while-councilor-jan-harder-and-development-consultant-jack-stirling-keep-it-in-the-family.

———. 2021. "Property Standards." Unpublished manuscript.

Rodimon, Sarah. 2018. "'We Have the Law, We Need the Access!': Activism, Access and the Social Organization of Abortion in New Brunswick." PhD diss., Carleton University.

Rolnik, Raquel. 2013. "Late Neoliberalism: The Financialization of Homeownership and Housing Rights." *International Journal of Urban and Regional Research* 37, 3.

Rucks-Ahidiana, Zawadi. 2021. "Theorizing Gentrification as a Process of Racial Capitalism." *City & Community* 21, 3. https://doi.org/10.1177/15356841211054790.

Russell, Hazen A.J., Gregory R. Brooks, and Don I. Cummings (eds.). 2011. *Deglacial History of the Champlain Sea Basin and Implications for Urbanization.* Ottawa: Geological Survey of Canada. https://doi.org/10.4095/289555.

Rutland, Ted. 2010. "The Financialization of Urban Redevelopment." *Geography Compass* 4, 8. https://doi.org/10.1111/j.1749-8198.2010.00348.x.

Rybczynski, Witold. 1987. *Home: A Short History of an Idea.* Reprint ed. New York: Penguin Books.

Sassen, Saskia. 2014. *Expulsions: Brutality and Complexity in the Global Economy.*

Cambridge: Harvard University Press.

Sawyer, Malcolm. 2013. "What Is Financialization?" *International Journal of Political Economy* 42, 4.

Saxena, Astha. 2021. "Through the Looking Glass, into Battlefield Herongate: Reflecting on a Site in Crisis as an Outsider." Master's thesis, University of Ottawa. hdl.handle.net/10393/42824.

Shaw, Marc. 2017. "Timbercreek Builds Resort-Style Apartments in South Ottawa." *Renx.ca*, December 14. renx.ca/timbercreek-builds-resort-style-apartments-south-ottawa.

Sheldrick, Byron. 2014. *Blocking Public Participation: The Use of Strategic Litigation to Silence Political Expression*. Waterloo, ON: Wilfrid Laurier University Press.

Shilton, Jamie. 2021. Who Owns the City? Pension Fund Capitalism and the Parkdale Rent Strike. *Journal of Law and Social Policy* 35. digitalcommons.osgoode.yorku.ca/jlsp/vol35/iss1/1.

Simpson, Michael, and Jen Bagelman. 2018. "Decolonizing Urban Political Ecologies: The Production of Nature in Settler Colonial Cities." *Annals of the American Association of Geographers* 108, 2.

Slater, Tom. 2009. "Missing Marcuse: On Gentrification and Displacement." *City* 13, 2–3.

___. 2011. "Gentrification of the City." In *The New Blackwell Companion to the City*, edited by Gary Bridge and Sophie Watson. Malden, MA: Blackwell Publishing.

Smith, Dorothy E. 1987. *The Everyday World as Problematic: A Feminist Sociology*. Boston: Northeastern University Press.

___. 1990. *Texts, Facts, and Femininity: Exploring the Relations of Ruling*. London and New York: Routledge.

___. 2001. "Texts and the Ontology of Organizations and Institutions." *Studies in Cultures, Organizations and Societies* 7, 2. https://doi.org/10.1080/10245280108523557.

___. 2005. *Institutional Ethnography: A Sociology for People*. Toronto: AltaMira Press.

Smith, Dorothy E., and Susan Marie Turner (eds.). 2014. *Incorporating Texts into Institutional Ethnographies*. Toronto: University of Toronto Press.

Smith, George. 1990. "Political Activist as Ethnographer." *Social Problems* 37, 4.

Smith, Neil. 1996. "After Tompkins Square Park: Degentrification and the Revanchist City." In *Re-Presenting the City*, edited by Anthony D. King. London: Palgrave.

___. 1997. Social Justice and the New American Urbanism: The Revanchist City. In *The Urbanization of Injustice*, edited by Andy Merrifield and Erik Swyngedouw, New York: New York University Press.

Smith, Neil, and Peter Williams (eds.). 2013. *Gentrification of the City*. Routledge.

Soederberg, Susanne. 2018. "Evictions: A Global Capitalist Phenomenon." *Development and Change* 49, 2. https://doi.org/10.1111/dech.12383.

___. 2021. *Urban Displacements: Governing Surplus and Survival in Global Capitalism*. London and New York: Routledge.

Statistics Canada. 2017. "Census Profile, 2016 Census." 5050007.02 [Census tract], Ontario and Ottawa, CDR [Census division], Ontario (table). Statistics Canada Catalogue no. 98-316-X2016001. www12.statcan.gc.ca/census-recensement/2016/dp-pd/prof/index.cfm?Lang=E.

Stein, Samuel. 2019. *Capital City: Gentrification and the Real Estate State*. London: Verso.

Story, Brett. 2019. *Prison Land: Mapping Carceral Power across Neoliberal America*. Minneapolis: University of Minnesota Press.

Surtees, R.J. 1983. "Indian Land Surrenders in Ontario 1763–1867." Ottawa: Indian and Northern Affairs Canada.

Suttor, Greg. 2009. *Rental Paths from Postwar to Present: Canada Compared*. Toronto: Cities Centre, University of Toronto.

———. 2016. *Still Renovating: A History of Canadian Social Housing Policy*. Montreal-Kingston: McGill-Queen's University Press.

Tedesco, Delacey, and Jen Bagelman. 2017. "The 'Missing' Politics of Whiteness and Rightful Presence in the Settler Colonial City." *Millennium* 45, 3.

Teresa, Benjamin F. 2019. "New Dynamics of Rent Gap Formation in New York City Rent-Regulated Housing: Privatization, Financialization, and Uneven Development." *Urban Geography* 40, 10. https://doi.org/10.1080/02723638.2019.1603556.

Thielen-Wilson, Leslie. 2018. "Feeling Property: Settler Violence in the Time of Reconciliation." *Canadian Journal of Women and the Law* 30, 3.

Thobani, Sunera. 2007. *Exalted Subjects: Studies in the Making of Race and Nation in Canada*. Toronto: University of Toronto Press.

Threndyle, Randy. 2009. "Timbercreek Finds Value in Overlooked Properties." *Canadian Apartment Magazine* 5, 6. issuu.com/rick11/docs/cam_feb09_issue.

Thurber, Amie, and Amy Krings. 2021. "Gentrification." In *Encyclopedia of Social Work*. https://doi.org/10.1093/acrefore/9780199975839.013.1413.

Timbercreek Asset Management. 2016a. "Heron Gate 7 Design Brief." July webcast. ottawa.ca/plan/All_Image%20Referencing_Site%20Plan%20Application_Image%20Reference_July%202016%20-%20D07-12-16-0104%20-%20Design%20Brief.PDF.

———. 2016b. "HG7 Planning Rationale Report & Design Brief." November.

———. 2019a. "Real Estate: Open a Window to Global Investing Opportunities." Brochure.

———. 2019b. "About Us." timbercreek.com/about-us.

———. 2019c. "Investment Vehicles." timbercreek.com/strategies/investment-vehicles.

———. 2019d. "Investment Strategies." timbercreek.com/strategies/investment-strategies.

———. 2019e. "Private Equity." timbercreek.com/strategies/private-equity.

———. 2019f. "Registrations Open for Timbercreek's New Vista Local Development." August 9. newswire.ca/news-releases/registrations-open-for-timbercreek-s-new-vista-local-development-860848340.html.

———. 2019g. "Heron Gate Master Plan, Ottawa: Public Open House (February 11)." Presentation slides.

———. 2019h. "Response to an Application under Section 34 of the *Human Rights Code*." September 27.

———. 2020. "Timbercreek Announces Rebranding of Business Units." *Cision*, November 9. newswire.ca/news-releases/timbercreek-announces-rebranding-of-business-units-890426394.html.

Timbercreek Communities. 2018a. "Re: Important Notice About Your Tenancy." Document distributed to tenants at a meeting held by Timbercreek on May 7, 2018.

———. 2018b. "Know Your Rights." Poster.

Timbercreek Investment Management Inc. 2019. "Timbercreek 2019 Global Real Estate Securities Outlook Report: Global REITs Expected to Deliver 9-10 Percent Returns." *GlobeNewswire*, January 10. globenewswire.com/news-release/2019/01/10/1686101/0/en/Timbercreek-2019-Global-Real-Estate-Securities-Outlook-Report-Global-REITs-Expected-to-Deliver-9-10-Percent-Returns.html.

Toews, Owen. 2018. *Stolen City: Racial Capitalism and the Making of Winnipeg*. Winnipeg: ARP Books.

Tolfo, Giuseppe, and Brian Doucet. 2022. "Livability for Whom? Planning for Livability and the Gentrification of Memory in Vancouver." *Cities* 123.

Tomiak, Julie. 2011. "Indigenous Self-Determination, Neoliberalization, and the Right to the City: Rescaling Aboriginal Governance in Ottawa and Winnipeg." PhD diss., Carleton University.

Tomiak, Julie, Tyler McCreary, David Hugill, Robert Henry, and Heather Dorries. 2019. "Introduction: Settler City Limits." In *Settler City Limits: Indigenous Resurgence and Colonial Violence in the Urban Prairies*, edited by Heather Dorries, Robert Henry, David Hugill, Tyler McCreary, and Julie Tomiak. Winnipeg: University of Manitoba Press.

Tonkiss, Fran. 2013. *Cities by Design: The Social Life of Urban Form*. Cambridge, UK: Polity Press.

Tranjan, Ricardo. 2023. *The Tenant Class*. Toronto, ON: Between the Lines.

Triece, Mary E. 2016. *Urban Renewal and Resistance: Race, Space, and the City in the Late Twentieth to the Early Twenty-First Century*. Lanham, MD: Lexington Books.

Tsourounis, Michael. 2016. "Newly Built Apartment Projects from Conception to Stabilization." Timbercreek Asset Management presentation slides, Canadian Apartment Investment Conference.

Tucker-Simmons, Daniel. 2018. "Re: Cease and Desist Letter dated 9 July 2018." July 19. In the author's possession.

Valiquette, Leo. 2019. "Timbercreek 'Trophy Markets' Focus Continues to Pay Off." *Renx.ca*, July 17. renx.ca/timbercreek-higher-end-trophy-markets-focus-2b-aum/.

Vanderheiden, Elisabeth. 2021. "The Terror of Being Judged: Public Shaming as Resource and Strategic Tool." In *Shame 4.0*, edited by Claude-Hélène Mayer, Elisabeth Vanderheiden, and Paul T. P. Wong. Cham, Switzerland: Springer.

Veracini, Lorenzo. 2015. *The Settler Colonial Present*. New York: Palgrave Macmillan.

Victoria Competition and Efficiency Commission. 2008. "A State of Liveability: An Inquiry into Enhancing Victoria's Liveability." Melbourne, Australia: Department of Treasury and Finance.

Vista Local. 2019. "Vista Local." vistalocal.com.

___. 2021. "Vista Local: The Community." vistalocal.com/community.

Walcott, Rinaldo. 2021. *On Property: Policing, Prisons, and the Call for Abolition*. Windsor, ON: Biblioasis.

___. 2023. Longing Across the Black Diaspora: Love, Being and the Door of No Return. TOPIA: *Canadian Journal of Cultural Studies* 46. https://doi.org/10.3138/topia-2022-0035.

Walks, Alan, and Brian Clifford. 2015. "The Political Economy of Mortgage Securitization and the Neoliberalization of Housing Policy in Canada." *Environment and Planning A: Economy and Space* 47, 8. https://doi.org/10.1068/a130226p.

Walks, R. Alan, and Larry S. Bourne. 2006. "Ghettos in Canada's Cities? Racial Segregation, Ethnic Enclaves and Poverty Concentration in Canadian Urban Areas." *The Canadian Geographer* 50, 3. https://doi.org/10.1111/j.1541-0064.2006.00142.x.

Ward, Kevin. 2007. "Business Improvement Districts: Policy Origins, Mobile Policies and Urban Liveability." *Geography Compass* 1, 3. https://doi.org/10.1111/j.1749-8198.2007.00022.x.

Watt, Paul. 2018. "'This Pain of Moving, Moving, Moving:' Evictions, Displacement and Logics of Expulsion in London." *L'Année Sociologique* 68, 1.

Webber, Cole, and Ashleigh Doherty. 2021. "Staking Out Territory: District-Based Organizing in Toronto, Canada." *Radical Housing Journal* 3, 1.

West Broadway Tenants Committee. 2021. "Communities vs. Starlight." Online event, May 25.
White, Rob. 2005. "Stifling Environmental Dissent: On SLAPPs and Gunns." *Alternative Law Journal* 30, 6.
Wijburg, Gertjan, Manuel B. Aalbers, and Susanne Heeg. 2018. "The Financialisation of Rental Housing 2.0: Releasing Housing into the Privatised Mainstream of Capital Accumulation." *Antipode* 50, 4.
Wilcox, Don. 2018. "Zibi a $1.5B Development in the Heart of Ottawa, Gatineau." *Renx.ca*, September 6. renx.ca/zibi-1-5b-development-heart-ottawa-gatineau.
Willing, Jon. 2018. "Incumbent Councillor's Steakhouse Fundraiser Cancelled Because of 'Threats,' He Says." *Ottawa Citizen*, September 22. ottawacitizen.com/news/local-news/incumbent-councillors-steakhouse-fundraiser-cancelled-because-of-threats-he-says.
Willing, Jon, and Adam van der Zwan. 2018. "Landlord Demolishing More Homes in Heron Gate, Displacing 105 Residents." *Ottawa Citizen*, May 8. ottawacitizen.com/news/local-news/landlord-demolishing-more-homes-in-heron-gate-displacing-105-residents.
Wiseman, John, and Kathleen Brasher. 2008. "Community Wellbeing in an Unwell World." *Public Health Policy* 29, 3. https://doi.org/10.1057/jphp.2008.16.
Withers, A.J. 2019. "Mapping Ruling Relations through Homeless Organizing." PhD diss., York University.
Wong, Nathalie. 2019. "Real Estate Investment Firm Timbercreek Hunts Beyond Canada's Borders for Undervalued REITs." *Financial Post*, July 18. financialpost.com/real-estate/property-post/real-estate-investment-firm-timbercreek-hunts-beyond-canadas-borders-for-undervalued-reits.
Xia, Lily. 2020. "Immigrants' Sense of Belonging in Diverse Neighbourhoods and Everyday Spaces." Master's thesis, University of Ottawa. https://doi.org/10.20381/ruor-24763.
Yussuf et al. v. Timbercreek Asset Management et al. 2019. File #2019-36509. Human Right Tribunal of Ontario.

Municipal Freedom of Information and Protection of Privacy Act Disclosures

City of Ottawa 2018-621

City of Ottawa 2018-629

City of Ottawa 2018-632

City of Ottawa 2018-744

City of Ottawa 2020-424

City of Ottawa 2021-549

Ottawa Police Service 19-494

Index

above guideline increases (AGIs), rent, 60, 72, 114, 139
ACORN Ottawa, 42, 127, 131
activist scholarship, 38–9
affordability, housing,
    activism, 42–4, 114, 118–21, 148
    claims of, 118–19, 131–4
    displacement for, 49–50, 65, 114, 143
    elimination of, 12–13, 72, 140, 142
    Heron Gate residents', 49–50, 52, 131–5, 142–3, 148
    manufactured crises of, 4–6, 133–5
    marginalized populations and, 1–5, 131–4
    metrics, 118–19, 130–3, 135, 148
    Ottawa Official Plan, 30, 127–32
    redevelopment and, 8–9, 85, 92–4, 104, 132–3
African diasporic communities, 22, 48, 52, 130, 143
Algonquin Anishinaabe peoples, 1, 140
    occupation of Ottawa River watershed, 10, 14–15
    settler colonial division of, 16–18, 29, 66
Ali, Abdullahi, 61–4
Alta Vista, 66, 118
    demographics of, 8, 80, 94, 104
    Heron Gate alignment with, 86–8, 99, 104, 128
    residents' revanchism, 91–2, 125
Apartment Watch program, 56–7
asset management firms, investment by, 70–1, 75; *see also* Timbercreek Asset Management
August, Martine, 70–2, 80, 92, 97, 117

belonging, sense of, 27, 69
    in diaspora communities, 23, 54–6, 64
Bevington, Douglas, 38
Bizzarri, Ugo, 67, 74, 79

Black communities,
    domicide against, 29–31, 92
    erasure of, 19, 69, 80, 86, 90–3
    Heron Gate, 51–2, 87, 110
    racist perceptions of, 57–8, 91, 113
    societal violence against, 21, 78, 93
    in white settler culture, 20, 23, 69
Blomley, Nicholas, 18, 20
branding, 29, 66–7
    management of, 78–9, 122, 127, 133–4, 146
    re-, 2, 7, 86, 108, 114, 126
British settlers,
    approach to asylum seekers, 22–3
    colonial, 15–17, 23
Butler, Judith, 26
bylaw office, municipal, 127
    maintenance order enforcement, 41, 45, 61–4
    tenant interactions with, 61–2

Canadian Apartment Investment Conference, 71, 75
Canan, Penelope, 108
capitalism, 97
    functioning of, 6, 36, 124
    racial, 20–1, 27, 30–1
    rentier, 4, 69, 71, 122
CAPREIT, 71, 76
city council, Ottawa, 43
    Heron Gate redevelopment support, 9–10, 42, 117, 134–8
    integrity issues with, 138–40
    Official Plan Amendment, 117, 126, 130–2, 134–6
city-making, 19
    settler colonial, 14–15, 20
    see also urban development
City of Ottawa, 52
    claims of consultation, 38, 44, 91, 116–17, 138–9

coalition organizing targeting, 105, 134–5, 140–1
declaration of housing emergency, 5
freedom of information (FOI) requests to, 41, 45, 60–3, 91
New Official Plan of, *see* Official Plan, Ottawa
officials, *see* officials, city
planning department, *see* planning, urban
restrictions on affordability demands, 128–31
support for Heron Gate demolition, 9–10, 12, 38, 44, 79–84, 115
class,
gentrification impacts due to, 4, 23, 26, 33, 55
hierarchies, 5, 63, 67, 89, 92–4, 124
middle, 89, 92
property, 70, 82, 94
working, *see* working-class neighbourhoods
Cloutier, Jean, 83–5, 96, 116–19, 127
commodification,
capitalist, 6, 68
de-, 4–5, 68
housing, 4, 45, 69, 135–6
land/property, 21, 123, 136–7
Community Wellbeing Framework, 115–16, 119–24, 147–8
court proceedings,
corporate use of, 106, 108–11
documentation analysis of, 44–7, 112
human rights, 7, 94–5, 142–7
participation in, 42, 111–12
tenant use of small claims, 45, 65, 95, 111–13
crime, 123
Heron Gate associations with, 56–9, 107
media coverage of, 58–9, 109–10
prevention projects, 56–7
criminality,
associations with Blackness, 57–8
coalition activist associations with, 109–10, 112–14
representations of, 24, 57–9, 109–10

debt, real estate, 3, 73
defamation, allegations of, 106–7
flipping threats of, 111–12
demolition,
government support for, 44, 83–5, 115–16, 125–9, 142
Heron Gate, 6–9, 41, 50, 61, 93–6
intensification and, 75–7, 82–4, 140
justifications for, 48–9, 56, 59–64, 81, 96
marginalized populations in, 29–30, 80–2, 119, 148
redevelopment and, 8, 65, 75, 87
demoviction, 138
analysis of, 42, 46, 61, 131
city refusal to intervene in, 44–5, 125
court case on, 142–4, 147
as domicide, 4, 10, 85–8
Heron Gate, 6, 9, 42, 77, 114
logics of, 10, 86–8, 94–5, 100
mass, 4, 65, 88, 94, 142, 148
redevelopment after, 80–4, 114
survey, 101–4
developers, 41
control of land, 28–9, 76, 82, 116–17, 139–40
intensification by, 75, 82–3, 97
lack of affordability by, 130–5
landlord-, *see* landlord-developers
legal cases involving, 108, 113, 144–6
municipal ties to, 30–2, 48, 86–8, 127–8, 136–40
practices of domicide, 8, 31, 34, 148
redevelopment presentations by, 42–4, 91, 117, 130–1
social framework and, 119, 125–7, 136, 148
Timbercreek involvement with, 58, 73, 75
DIALOG (design company), 88, 117, 120, 138
diasporas, 7
genealogies of dispersion, 22–3
theorizing on, 22–3
tropes of, 58
unmaking of, 29–30
see also ethnoracial enclaves
dilapidation, conditions of, 24, 59–60, 64
direct action, 37, 39

direct engagement, 35, 38–40, 140
discourse, housing,
    ethnographic analysis of, 35–8, 110
    improvement, 9, 24, 30–4, 76–7, 86, 95
    liveability, 24–7, 30, 116, 123–4, 137
    manufactured crisis, 4–5
    ruling relations from, 8, 80, 88, 147
    settler colonial, 17, 19–20, 23
discrimination,
    employment, 53
    housing, 53–4, 142, 145–6
    racial, 133, 142–6
    structural, 48–50, 64, 103
displacement, 119
    contesting, 97–100, 105, 109, 114, 127, 140
    via gentrification, 4, 7–9, 31–3, 74, 80, 143–4
    Heron Gate residents', 24, 48–9, 53–5, 78, 145–6
    mass, 61, 82–4, 97–8, 118, 148
    racialized, 19, 25–8, 65–71, 93, 128–32
    settler colonial, 15, 18–21, 27, 76–7
    trend of, 90
disposability, 124
    of particular lives, 9, 26, 31, 33–4, 147
dispossession,
    domicide and, 23, 28–9
    ongoing settler colonial, 6, 18–21, 31–3
Dixon, Chris, 38
domicide,
    analysis of, 47, 51
    concepts and use of, 4, 24, 27–31
    developer practices of, 8, 30–4, 148
    government support for, 9, 12, 38, 44–6, 79–84, 115
    liveability discourse and, 28–34, 38, 94–5, 124–5, 147
    logics of, 10, 88
    racialized migrant, 22–3, 30, 33–4
    resistance to, 4, 39, 97–100, 109, 141, 147
    settler colonial, 16–17, 29, 31, 77–8
    violence of, 30, 85
Drimmer, Daniel, 2, 66–7

Ellis-Young, Margaret, 33
empire,
    British, 15, 23
    migrant flows and, 22–3
ethnography,
    analysis and practices of, 41, 44, 47
    institutional, 35–7, 40
    political activist, *see* political activist ethnography
ethnoracial enclaves, 48, 56, 140, 144
    Heron Gate Village as, 7–8, 22–3, 53–5, 142–3
European squatters/settlers, 49
    global imperialism of, 22
    Ottawa River watershed, 15–17
    property relations of, 16–17
evictions,
    developer modifications to, 128, 140–1, 148
    discourse on, 4, 25, 64, 81–7, 114, 118
    domicide through, 4, 10, 28–30
    government support for, 44–6, 61–2, 138
    intensification and, 75–8, 82–4, 140
    mass, 7–8, 55, 58–61, 93–8
    resistance to, 41–4, 51, 98–102, 105, 109
    rising rates of, 4, 43
    threat of, 42, 106
    urban renewal, 49, 77–80, 85, 132

Farha, Leilani, 65, 68, 99
financialization,
    documentation of, 43–5
    gentrification and, 70–2, 77–8, 94–5, 148
    landlord, 2, 8, 65, 78, 82
    real estate firm, 65–7, 70–1, 75, 108, 144–5
    of rental housing, 4, 31, 64, 67–70, 119
    tactics of, 6–7, 51, 62, 71–3, 89
    tenant struggles against, 97–100, 111
freedom of information (FOI)
    research, 43–4, 84, 91, 113, 137
    use in legal proceedings, 41, 45–6, 60–2

Gashan, Saido, 61–4
gender,
    financialized landlords and, 78
    gentrification impacts due to, 4, 23, 26
gentrification,
    actors, 32, 67, 75–6
    affordability and, 130–4
    displacement and, 4, 7–9, 31–3, 74, 80, 143–4
    domicide through, *see* domicide
    financialized, 70–2, 77–8, 94–7, 144–5, 148
    government support for, 9–10, 42, 83–6, 117, 134–8, 142
    intensification and, 74–7
    liveability discourse and, 24–7, 30–2, 116, 123–4, 137
    processes of, 7–8, 31–2, 50–1, 60, 128, 137
    revitalization and, 24, 30–3, 45, 70–4, 77, 80–2
    struggles against, 97–100, 111, 114
    targeted, 6, 30, 65
    vulnerability to, 68, 70, 78, 89–93, 146–7
    whiteness and, 31–3, 91–4
ghettoization, 24, 58–9, 92
Gómez-Palacio, Antonio, 117–19
grassroots organizing, 42, 97, 105, 111, 128, 140
Gréber, Jacques, 49
Greenbergs, Irving and Dan, 66

Harder, Jan, 41, 138–40
Hawley, Josh, 65–7, 83
Hazelview, 7, 45
    affordability, lack of, 131–5, 147–8
    MOU allowances for, 127–8, 131, 134
    narratives of, 77–8, 134, 145–6
    Timbercreek rebranding as, 2, 65, 78–9, 114, 126
Heatherington, 1, 7, 40, 45, 50, 98
Herongate (neighbourhood), 1, 6
Herongate Tenant Coalition,
    activist scholarship with, 39–47, 54, 137
    community meetings of, 42, 98–9, 109
    eviction organizing, 7, 60–1, 97–8, 104–11, 114–15, 139–42
    interviews with, 43, 87–9, 132
    legal proceedings, 47, 63, 107, 111–13, 146
    media campaigns by, 84, 98–104, 107–11
    research by, 45, 62, 65–7, 138
    Timbercreek versus, 63, 78, 101, 105–9, 144–8
Heron Gate Mall, 6–7
Heron Gate Village, 140
    affordability of housing in, 49–50, 52, 131–5, 142–3, 148
    attempted destruction of, 6, 65, 78, 94, 128
    Black communities in, 51–2, 87, 110
    bombing of, 84–5
    city support for demolition, 9, 38, 44, 79–84, 115
    criminality, associations with, 56–9, 107–10, 112–14
    demographic composition of, 7–8, 50–2, 88–9, 102–4
    displacement of, 24, 48–9, 53–5, 78, 145–6
    domicide in, 24, 28–34, 38, 94–5, 124–5, 147
    as ethnoracial enclave, 7–8, 22–3, 48, 53–6, 142–4
    revitalization narratives of, 31–3, 48–9, 81–2, 92–5, 120
    settler colonial history of, 14, 18–23
    sociocultural networks of, 7, 50, 53–9, 78, 143, 148
    Timbercreek neglect of, 41, 48, 60–1, 83–4, 144
    Timbercreek purchase and demolition of, 2, 7, 41
HG7 (Kanco Heron Gate-7), 81–2, 86–8
homelessness, 3, 5–6
homeownership,
    Alta Vista, 86–7, 91–2, 125
    focus on, 3–5, 68
    settler norm of, 3, 30, 125
housing crises, 113–14
    ignorance of root causes, 4–5, 30
    as manufactured, 4–6
housing justice, 113, 146
    activism, 5, 41, 43, 97
housing systems,

conflicts in, 4–5
inequalities in, 17–20, 69
tenant organizing versus, 5–6

immigrants, 4, 102
community displacement, 7, 30, 49, 55–6, 143–4, 148
in gentrification processes, 67, 77, 86, 132–3, 144–7
Heron Gate as neighbourhood for, 8, 51–5, 59, 104, 132
media stigmatization of, 58
settler colonial treatment of, 22, 30, 67, 144
Indigenous land,
expropriation of, 10, 16–18, 27–8, 140
homeownership on, 3, 65
as stolen, 9, 14–19, 67, 91
white settler profit from, 30, 65–7, 71
see also Algonquin Anishinaabe peoples; land; settler colonialism
Indigenous Peoples,
dispossession of, 6, 18–21, 31–3
settler colonial depictions of, 17, 20
white settler violence against, 23, 27
inequalities, 53
housing, 52, 97, 103–4
racial, 25, 69, 103–4, 137
spatial, 32, 69–70, 137
systemic, 25–6, 103–4
inner city,
councillors, 132, 135, 139
residents' displacement, 49, 81
intensification,
asset generation through, 69, 75, 82
financialization and, 62, 75–7, 95
Heron Gate, 83, 125
Ottawa New Official Plan on, 9–10, 30, 76, 147
settler colonial, 14–15
interviews,
City of Ottawa employee, 9, 43, 117, 136–7
Herongate Tenant Coalition organizer, 54–6, 87, 89, 99, 106
Heron Gate resident, 8, 43, 59, 90–3, 110
political activist ethnography and, 35, 43–4, 47

Timbercreek official, 74, 87
investors, real estate, 111,
community wellbeing versus, 122, 135–6
company messaging for, 45–6, 72, 79, 114
focus on returns, 65, 77, 79, 94–5
institutional versus individual actors, 70–2, 87–8
intensification and, 72–7
legislation favouring, 68–9, 107–8
municipal governments and, 136–7
see also real estate investment
Iraqi diasporic communities, 52–3
Islamophobia, 78, 144

Kichi Sibi watershed, 15–16; see also Algonquin Anishinaabe peoples
Kinsman, Gary, 37
knowledge production, 46
movement-relevant theory and, 35, 38–40
social movement, 8, 36–9, 98, 146
Krempulec, Colleen, 132–3

land,
demoviction from, 9, 88
intensification and, 75–6, 81–2
investor thirst for, 69, 76, 94
ownership, 6, 18, 66, 116
settler colonial reconfiguration of, 16–21, 29–33, 94
use policies, 5, 10, 32, 50, 125–6, 136–7
see also Indigenous land
landlord-developers,
activist scholarship against, 39–40, 42–7
claims of consultation, 38, 42–4, 87, 91–2, 116–17, 130
government support for, 44, 60–2, 68–9, 115, 136–8
Heron Gate ownership, 6–8, 73
see also Timbercreek Asset Management
landlords, 85
challenging, 39, 44–5, 105, 108–12, 130, 140–5
developer, see landlord-developers
financialized, 2, 8, 65, 70–1, 77–8, 82

loopholes for, 62, 70–2, 82–3, 115–16, 125–8
narratives of, 56–60, 81–2, 86–9, 117–18
neglect, *see* neglect, landlords'
predatory, 51, 74, 101, 110–11, 146
resistance to, 2, 98–101, 146
responses to tenant organizing, 99–101, 105, 108–114, 130–4, 148
Landlord and Tenant Board, 99
activism against, 42
AGI applications to, 60, 72
Land Registry Office documents, 65–6
LeBreton Flats, 48–50, 53
Ledbury–Heron Gate–Ridgemont, 49–51
legal proceedings, 99
documentation of, 42, 46, 110–11
human rights, 7, 94–5, 142–7
lawsuits against Timbercreek, 62–4, 78, 111–13
settler colonial property relations, 17–21, 83, 94–5, 146
Timbercreek repression through, 104–9
see also court proceedings
Leiper, Jeff, 134–5
Leveller, The, 39, 41, 66, 139
liveability, urban,
Alta Vista harmonization and, 80, 85–8, 99, 104, 128
designation of, 10, 24, 45, 76
discourse of, 25–7, 30, 33–4, 115–18, 120–4, 137
domicide accompanying, 24, 31–4, 38, 94–5, 128, 147–8
Heron Gate community, 6, 50, 53–6, 88–90, 124–5, 147–8
indexes of, 25, 46, 50
lack of, 26, 33, 56, 61–3, 130, 147
New Official Plan on, 9–10, 19, 24, 31, 76–7, 147–8
precarity and, 3–6, 26, 30–1, 56, 147
production of, 26–7, 30–1, 77–8, 128
rankings of, 25
revitalization and, *see* revitalization
varying meanings of, 24–7, 33–4, 43, 50
wellbeing and, 115–18, 120–4, 147–8

whitespace and, 8, 26–7, 31–4, 88
Loubser, John, 62–3
lower-income households,
displacement of, 4–5, 9, 65–7, 81, 132, 143–7
Heron Gate, 67, 102–4, 148
in rental market, 1–3, 49–51, 68, 131
revanchist attitudes towards, 91–2, 94
revitalization targeting, 30, 33, 45, 70–4, 77

maintenance, property,
deteriorating, 2–3, 60–1, 81, 130
developer refusal of, 48, 60–4, 66, 83
lack of enforcement of, 41, 44–5, 61
requests/orders, 60, 63–4, 104–5
tenant responsibility for, 60–3, 75, 130
see also neglect, landlords'
market, housing,
affordability, *see* affordability, housing
financialization of, 64, 68–9, 77–8
multiresidential properties, *see* multiresidential property market
out of reach, 3, 31, 82, 126
policies, 5, 30, 133–4
real estate investment firms in, 2, 64, 69–75
McNeilly, Kathryn, 26
media, 5, 84–5
Herongate Tenant Coalition campaigns, 78, 98, 109
Heron Gate coverage, 7, 45–6, 56–9
independent work in, 39, 41, 43–4
Timbercreek concern for, 82, 87, 110–14
see also social media
memorandum of understanding (MOU),
details of, 12, 79, 119, 126–8, 130–5
negotiating, 12, 115, 127–8, 136
opposition to, 127, 134–6, 140
Menard, Shawn, 132–4
Mensah, Joseph, 8, 51, 53
Middle Eastern diasporic communities, 22, 48, 52
migrants,
communities of, 22, 52–3, 55, 143
domicide against, 17, 21–3, 30–3
Mi'kma'ki, 1–2

Minto Construction, 66
Momentum Planning and Communications, 100–1
  public relations work, 81–2, 91–3
movement-relevant theory, 35, 38–40
multiresidential property market, 71, 74, 79
municipal-developer nexus, 31, 62-3, 83, 88, 136–7, 146
municipal governance,
  developer ties to, 31, 62, 83, 136–7, 146
  discourses of, 26, 146–7
  domicide, support for, 9, 46, 83
  mechanisms of gentrification, 9, 31, 62
  ruling relations of, 36, 40, 76, 124
  settler colonial, 16, 19
Muslim population, 87, 110
Mustang Equities, Inc., 67, 79, 143

neglect, landlords', 109
  of Herongate, 41, 48, 60–1, 83–4, 144
  justification for demolition, 48, 60–1, 74, 80–1
  stigmatization due to, 50, 56, 59–61
  systemic and purposeful, 2, 44, 62–4, 82–5, 130
  transferring blame for, 60, 63
  see also maintenance, property
Neighbourhood Watch program, 56–7
neoliberalism,
  housing policy under, 4, 72
  governance, 57, 68, 124
Nepali diasporic communities, 7, 52–4, 126–7

Official Plan, Ottawa,
  on affordability, 30, 127–32
  Amendment, 79, 115–16, 119, 125–6, 140–1
  amendment process, 130–6, 138–40
  claims of consultation, 116–17
  content of, 9–10, 30, 76, 123
  documentation of, 42, 44–6
  gentrification, support for, 30, 76, 79, 84–6, 142
  improvement discourse in, 19, 30, 76
  on intensification, 9–10, 30, 76, 147
  on liveability, 9–10, 19, 24, 31, 76–7, 147–8
officials, city, 9, 109, 117, 146
  links with developers, 41, 83, 105, 136–9
  liveability discourse of, 25–6, 56, 83–5, 133, 140
  practices of domicide, 8, 10, 31, 34, 78, 147
Ontario, 125
  activist movements in, 37, 43, 97
  British colonization of, 15–18
  housing deregulation in, 62, 68, 72
  landlord organizing in, 112–14, 144
  rental guidelines for, see Residential Tenancies Act
Ontario Human Rights Tribunal case, 7, 13, 53, 113, 142–7; see also tribunals, landlord-tenant
Ottawa,
  as Algonquin Anishinaabe territory, see Algonquin Anishinaabe peoples
  City of, see City of Ottawa
  councillors, see city council, Ottawa
  official plan of, see Official Plan, Ottawa
ownership, 61, 125, 146
  changes in housing complex, 2, 60, 66–7, 79
  Heron Gate, 6, 65–8, 79, 94
  house, see homeownership
  real estate investment firm, 2, 66–7, 70–2, 75–8, 127
  settler colonial systems of, 17–18, 21, 66, 94

Parkdale, 2, 91
Parkdale Organize, 97–8
planners, urban, 14
  developers, links to, 31, 41, 63, 88, 136–9
  domicide by, 8, 78–82, 147
  redevelopment projects and, 31–2, 49–50, 66, 80, 86–91
planning, urban, 29
  documents, 10, 43–5, 81–2, 91; see also Official Plan, Ottawa
  on Hazelview, 132–4, 147

on Heron Gate, 31, 42, 78–83, 86–91, 147
liveability in, 25–6, 88–90, 116–17
social framework in, 126–8, 131
trends, 50, 76
police,
bombing of Heron Gate, 84–5
criminalization of activists by, 109–13
increased presence of, 57–8, 93
Timbercreek connections with, 112–13
policy, housing, 44, 58
developer connections to, 72, 97, 128–9
federal influence in, 3, 28, 32, 50
liveability in, 9, 25–6
municipal, 5–6, 10, 14, 76–7
neoliberal, 4, 68–9
provincial legislation and, 68, 125–6
settler colonial, 28, 67
political activist ethnography, conceptualization of, 8, 37
data collection and analysis, 40–7
research design, 35, 38–40, 140
see also ethnography
Porteous, Douglas, 24, 28
poverty, 37
gentrification's displacement of, 24, 32, 51, 70, 76, 130–2
in Heron Gate, 48, 50, 53, 58, 64, 102–4
racialized, 24, 48, 56, 80–1
power,
activist analyses of, 38, 61, 116
building tenant, 5–6, 99, 105–7, 137
institutional relations of, 15, 36–7, 111
purchasing, 9, 60, 68, 70
settler colonial systems of, 15, 19, 22–3
precarity,
housing, 3, 5, 42, 57, 93
perpetuation of, 30, 73–4, 93
unequal distribution of, 5, 26, 147
Pring, George William, 108
private property,
ruling relations of, 9, 140
settler regimes of, 16–18, 21
profit,

domicide and, 133–5, 137, 140
focus on maximizing, 10, 25, 65–6, 70–1, 108
housing for, 5, 30, 73, 76
liveability discourse and, 30, 33, 83, 122, 148
"squeezing," 62, 71–2, 75, 82, 135
property relations, 124, 147
ethnoracial enclaves and, 23, 137
municipal-developer, 31, 63, 136
private, see private property
racial logics of, 8–9, 18–21, 24, 94
settler colonial, 16–21, 28–33, 65–7, 77–8
Property Standards and License Appeals Committee (PSLAC), 41, 62–3
public relations, developer, 81–2, 84, 114
consultation, 91, 112, 117, 140
public shaming, tactic of, 105, 108, 111, 146

Quebec, 2, 16

race,
gentrification impacts due to, 4, 22, 26, 92, 94
housing census incorporating, 51–2
settler colonial logics of, 19, 26, 78
racialization,
exclusion and, 23
property relations, 20–4, 27
racialized communities,
domicide in, 9–10, 22–3, 28–34, 94–5, 124–5, 147
enclaves for, see ethnoracial enclaves
eviction of, 8–9, 19, 25–8, 65–71, 93, 128–32
Heron Gate, 7–8, 22, 50–2, 88–9, 102–4
human rights case involving, 7, 94–5, 142–7
inequalities facing, see inequalities
poverty in, 24, 48, 50, 53, 56, 80–1, 102–4
reconfiguring identities in, 23
sociocultural networks of, 7, 50, 53–9, 78, 143, 148
unmaking of, 17, 27–9, 77, 85, 147
racism, 92

in media representations, 57–9
structural, 52–3, 76–8, 103–4
Razack, Sherene, 20, 27
real estate development,
profitmaking in, 5, 64, 79, 82
ruling relations in, 40, 92, 94, 136–7
settler colonial, 20–1, 30–3
use of SLAPPs, 107–9
real estate investment, 111
conferences, 42, 44, 77
firms, *see* real estate investment firms
in housing market, 2, 31, 64, 66–75, 144–5
see also investors, real estate
real estate investment firms,
analysis of, 43, 45, 51, 69–70, 119
financialized, 45, 65–71, 75, 108, 144–5
ownership, 2, 66–7, 70–2, 75–8, 127
tenant organizing against, 6, 97–8, 111, 119, 144–6
Timbercreek as, 6–7, 45, 78–9, 144
real estate investment trusts (REITs), 98
housing market takeover, 70–4, 76
property redistribution among, 66–7, 79
redevelopment, housing,
affordability and, 8–9, 85, 92–4, 104, 132–3
demovictions and, 8, 65, 75, 80–4, 87, 114
developer presentations on, 42–4, 91, 117, 130–1
discourses of, 28–34, 38, 76–7, 94–5, 124–5, 147
municipal support for, 9–10, 42, 117, 134–8
urban planning and, 31–2, 49–50, 66, 80, 86–91
refugees, 22–3, 85
Heron Gate as home to, 7, 51–6, 104
marginalization of, 30, 58
regeneration, 30, 76; *see also* intensification
Regent Park (Toronto), 80–1, 91
rent, 96, 143
above guideline increase (AGI) to, 60, 72, 114, 139
affordable, *see* affordability, housing

below-market, 30, 77, 82, 118–19, 126, 142
controls, removal of, 68
"premium," 78, 82, 87–8, 92–4, 118, 142
profitmaking from, 71–3, 82, 87
spiralling, 3, 128, 131–2
strike, 2, 97–8, 114
rental housing,
deregulation of, 62, 64, 68–9, 72
financialization of, 4, 31, 44, 64–71, 97–8, 148
Heron Gate residents in, 52
inadequacy of, 4, 30, 144–5
insecurity, 4, 21, 74, 93, 97
unaffordability of, 4, 44, 119–20, 133–4
repositioning buildings, 64, 72, 74, 82
Residential Tenancies Act, 62, 83
revitalization,
concept of, 33, 50, 80
demolition for, 8, 48–50, 65, 75, 85–7
discourse of, 9, 24–6, 31–3, 81–2, 92–5, 120
domicide through, 4, 28–34, 38, 94–5, 124–5, 147
gentrification through, 24, 30–3, 45, 70–4, 77, 80–2
municipal focus on, 9–10, 38, 44, 49, 79–84, 115
New Official Plan, 10, 19, 30, 76, 79, 84–6, 142
targeting low-income communities, 30, 33, 45, 70–4, 77
white settler notions of, 10, 27–32
see also vitality
rights, 14
case for human, 13, 94, 135, 140–6
education on, 99–101, 105–6, 109
housing, 2, 43, 135, 145
land/property, 16, 21, 137, 146
organizations, 41–2, 68, 113
to return, 7, 13, 83, 127, 143, 147
tenant, 2, 98–101, 105, 109, 113
Rodimon, Sarah, 38, 46
Rogers, Greg, 87, 91–2, 96, 117–19, 133
ruling relations, 35, 88, 118
activist disruption of, 8–9, 37–40, 46–7, 111, 140

concept of, 8, 36, 40, 44
Russo, Corrado, 73–5

seniors, housing for, 3, 73
settler cities, 9, 71, 124
    relations of, 19–20
settler colonialism,
    in Algonquin homeland, 10, 15–18
    creation and reproduction of, 14, 19, 26–33, 137
    dispossession, *see* dispossession
    property relations of, 16, 36, 65–9, 77–8, 146
    racial capitalism and, 20–1, 23–4, 94
settler colonial urbanism, 14, 20–1, 30–3, 137, 140, 145–8
settler-emplaced subjectivity, 22–4, 27, 31–4, 124
settler states, 4, 15–16, 19
    elimination of racialized spaces, 21–3
SLAPPs, 107–9
slums, clearance of, 48–9, 76
Smith, Dorothy, 36
Smith, George, 37
Smith, Neil, 92
Smith, Sandra E., 24, 28
Snedden, Lee Ann, 134–5, 137
social framework, 136, 148
    lack of affordability in, 119, 120, 127–35, 140
    negotiations for, 115–16, 126–31
    principles for, 120, 125, 127
    Timbercreek presentations on, 119–20, 126–31
    see also Community Wellbeing Framework
social media, 59
    Herongate Tenant Coalition campaigns, 46, 78, 105–6, 140
    Twitter, 105–6, 109–12
    use of, 95, 98, 146
social movements, 24, 44
    legal system engagement, 107, 111
    research with, 8, 35, 37–9, 40
    tactics of, 105–6, 146
    theorization of, *see* movement-relevant theory
Somali people,
    diasporic communities of, 52–5, 126

evictions of, 8, 102, 118
organizing of, 7, 98–9, 143
stigmatization of, 58–9, 89, 92–3
South Asian diasporic communities, 22, 48, 52–3, 143
Stein, Samuel, 32, 50, 136
stigmatization, 1, 110
    neighbourhood, 50, 57–8, 63–4
    racial, 21, 48–9, 56–9, 80–1
Stirling Group, 91–2, 138–40
suburbs, 7, 81
    resident forced relocation to, 49, 143
    sprawl of, 68
sustainability, environmental,
    neighbourhood marketing, 29–30, 77, 85–7, 118, 128
    policymaking emphasis on, 121, 123, 125
Syrian diasporic communities, 52–3

Tamblyn, Blair, 67, 74, 79
Tenant Protection Act, 62, 68, 72; *see also* Residential Tenancies Act
tenants, 122, 127, 133
    activist research with, 35, 39–44, 46–7
    building power of, 5–6, 105–7, 137, 142–7
    landlords versus, 38, 60, 77–8, 82, 107–15
    organizing, 39–44, 46, 53, 96–106, 114
    precarity of, *see* precarity
    replacing lower-income, 4, 9–10, 49, 65, 70–7, 92, 124
    responsibility for maintenance, 60–3, 75, 130
    rights for, *see* rights
    see also Herongate Tenant Coalition
textual record collection, 35–7, 40, 44–6, 118, 130
Timbercreek Asset Management,
    as gentrification agent, 4, 7–9, 31–3, 73–6
    Herongate Tenant Coalition versus, 63, 78, 101, 105–9, 144–8
    lawsuits against, 62–4, 78, 111–13
    mandate and vision, 72–3, 87, 92, 116–18, 125

neglect of properties, 41, 48, 60–1, 83–4, 144
predatory strategies of, 51, 73–9, 101, 104–11, 146
property sales to, 2, 6–8, 41, 73
see also Hazelview
Tomiak, Julie, 18, 21
Toronto,
financialization of housing in, 69–70, 80
housing rights activism in, 2
landlord loopholes in, 62, 133
TransGlobe, 67
housing decay, 2, 60, 66
property sales to, 2, 57, 66
tribunals, landlord-tenant, 146; see also Ontario Human Rights Tribunal case
Tucker-Simmons, Daniel, 8, 51, 53, 99, 106

United Nations Human Rights Council report, 68–9
Upper Canada, 16–17
urban development, 18–19, 125
activist research on, 42–3, 45–7
domicide through, 4, 8, 30, 38
improvement discourses, 19, 24–30, 77, 85–6, 137
racialized communities as obstructing, 16, 21
settler colonial logics of, 10, 14–15, 27–8, 31, 145

Valiquet, Suzanne, 91, 112–13
viability, building,
notions of, 85–6
profitability and lack of, 33, 60, 63, 83, 96, 148
violence, 107
domicidal, 29–31, 33, 69, 81–2
fleeing, 22
ongoing Western, 22
police, 85, 113
stereotypes of racialized, 58, 110
white settler colonial, 19–21, 26–7
Vista Local,
affordability, lack of, 94–5, 114, 118, 134
demovictions for, 8, 80, 92, 126, 138
marketing for, 77, 80, 85, 87–90, 127
vitality,
liveability focus on settler, 10, 30–3, 120
neighbourhood destruction and, 33, 50, 86, 128

Walks, Alan, 69–72
wealth, 68, 104
accumulation, focus on, 27, 69–70, 94
landlord-developer, 2, 66, 108, 111
white settler colonial, 16, 22, 27, 86–7, 91–3
whiteness,
domicide and, 16–17, 22–3, 29–31, 33–4, 77–8
gentrification and, 31–3, 91–4, 140
Heron Gate alignment with, 8, 48–9, 81–2, 92–5, 120
land settlement and, 15, 18–20, 23, 27, 65–9, 77–8
liveability discourse and, 8–9, 26–7, 31–4, 88
logics and relations of, 19–22, 29, 33, 57
privilege, defence of, 39, 67, 88–90, 92–4
racialized community replacement with, 10, 22–4, 29, 86, 142–8
resistance to, 23, 54–6, 64
violence of, 19–21, 23, 26–7
whitespace, 33, 86, 140
production of, 19, 26–7, 29–31
white supremacy, 20–2, 33, 59, 78
women, 58
activism of, 36, 43, 99, 110
community supports for, 29, 55, 78
domicide impacts on, 29, 78, 102
working-class neighbourhoods,
demolition of, 29, 48–50, 55
gentrification of, 32, 67, 80, 128, 147
organizing of, 99, 148
see also class

Xia, Lily, 8, 50, 53–4, 59, 81

Zibi, 28–9